Mathematical Induction 101

With 101 Practice Problems

Duc Van Khanh Tran

Illustrations by **Vy Nguyen Tong**

Table of Contents

About the Author

Hi! My name is Duc Van Khanh Tran, and I am a Vietnamese undergraduate at the University of Texas at Austin majoring in Mathematics.

When I was in elementary school, I studied at Morinosato Elementary School in Kanazawa, Ishikawa province, Japan for about two years. For middle school education, I studied at Le Quy Don Middle School in Ho Chi Minh City, Vietnam.

When I was in 9th grade, I came to the USA and started my high school study at Brentwood Christian School in Austin, Texas. The classes in my high school were more relaxing and less stressful than the classes in my middle school, so I began understanding and enjoying the things I studied at school. After a while, I realized that I enjoyed math the most out of all the subjects. Also, there was a math team in my high school, and participating in the math team made me love mathematics even more. That is how I "fell in love" with mathematics.

When I was in 11th grade, I came across *@daily_math_*, a very popular math account on Instagram. Inspired by the *@daily_math_* page, I also created a math account on Instagram called *@dvkt_math* with about 33,000 followers currently. Other than sharing mathematics knowledge by the means of a math page on Instagram, I decided to also write math books. Before writing this book, I also wrote three other books:

- *An Introduction to Calculus: With Hyperbolic Functions, Limits, Derivatives, and More* (May 2021);

- *Integrals and Sums Fiesta: An Integral Part of a Math Enthusiast's Life* (November 2021);

- *Basic Linear Algebra: An Introduction with an Intuitive Approach* (May 2022).

Preface

Mathematical induction is one of the common methods of proof in mathematics. It is a method to prove a statement about natural numbers n by figuring out the pattern from some particular cases, such as $n = 1$, $n = 2$, $n = 3$, etc. The method of mathematical induction is explained further in the "Prerequisites" chapter at the beginning of the book.

After the "Prerequisites" chapter, there are 101 problems for the readers to practice with mathematical induction. This book aims to introduce the method of mathematical induction to those who are unfamiliar with it while also providing some interesting mathematical induction problems to those who are already familiar with it. So, this book would be suitable for students who want to learn about mathematical induction or who are looking for more problems to practice with mathematical induction.

Most of these problems are either created by myself or collected from online sources such as Math Stack Exchange, Instagram, Twitter, etc. A few problems are taken from or inspired by problems in the book *The Cauchy-Schwarz Master Class* by J. Michael Steele and past preliminary exams of Singaporean junior colleges. (All of the solutions are written by myself and not taken from others.)

While the practice problems cover a wide variety of topics (as you can see at the top of each problem), most problems in this book only require the prerequisite knowledge of high school Algebra and introductory Calculus. A few problems also require some familiarity with the basics of Linear Algebra and Set Theory. There might be some unfamiliar concepts to those who have only taken high school Algebra and introductory Calculus, but some important definitions required to understand the problems are provided in each problem.

Lastly, please note that since we are only proving using mathematical induction in this book and the solutions are written by myself alone, some problems in this book could have other ways to prove, either with or without using mathematical induction. The readers are encouraged to find other possible methods of proof to improve their proof skills.

I hope you enjoy the variety of mathematical induction problems awaiting ahead!

Duc Van Khanh Tran
Texas, USA, 2023

Prerequisites
Mathematical Induction

Mathematical induction is a proof method that can be used to prove a statement about natural numbers $n \in \mathbb{N}$. To clarify the notations, in this book $\mathbb{N}$ does not include 0, and the letters k, m, and n always represent a natural number.

Before discussing the mathematical induction method itself, let us discuss why it is called "mathematical induction." There are mainly two types of proof: deductive proof and inductive proof. A deductive proof is where we prove a general statement directly by using axioms, definitions, proven theorems, etc. An inductive proof is where we prove a general statement by proving some specific cases first and then somehow relying on those specific cases to show that the statement is true for all cases. As you will see, mathematical induction relies on some specific cases called base cases.

In this book we will focus on two most common types of mathematical induction called weak induction and strong induction.

Weak Induction

To prove a statement $S(n)$ for all natural numbers n, we prove the following.

1. $S(1)$ is true;

2. If $S(k)$ is true for some $k \geq 1$, then $S(k+1)$ is also true.

Now let us discuss how this method can prove $S(n)$ for all $n \in \mathbb{N}$. First, we prove that the base case $S(1)$ is true. Then, we prove the induction step, which is $S(k+1)$ is true if $S(k)$ is true. By applying the induction step to $S(1)$, we can conclude that $S(2)$ is true. Then, we can also conclude that $S(3)$ is true because $S(2)$ is true, $S(4)$ is true because $S(3)$ is true, $S(5)$ is true because $S(4)$ is true, and so on. Thus, this method can prove that $S(n)$ is true for all $n \in \mathbb{N}$.

This method of weak induction can be modified slightly to prove a statement $S(n)$ for all $n \geq a$ for some natural number $a > 1$.

Weak Induction (Modified)

To prove a statement $S(n)$ for all $n \geq a$, we prove the following.

1. $S(a)$ is true;

2. If $S(k)$ is true for some $k \geq a$, then $S(k + 1)$ is also true.

Since we are proving a statement $S(n)$ is true for all $n \geq a$, the base case is $S(a)$ instead of $S(1)$. Then, applying the induction step, we can conclude that $S(a+1)$, $S(a+2)$, $S(a+3)$, etc. are true. Thus, we can prove that $S(n)$ is true for all $n \geq a$.

Strong Induction

To prove a statement $S(n)$ for all natural numbers n, we prove the following.

1. $S(1), S(2), \cdots , S(\ell)$ are true for some natural number ℓ;

2. If $S(m)$ is true for all $1 \leq m \leq k$ for some $k \geq \ell$, then $S(k + 1)$ is also true.

Now how does this method prove that $S(n)$ is true for all $n \in \mathbb{N}$? First, we prove the base cases $S(1), S(2), \cdots , S(\ell)$. Then, we prove the induction step, which is $S(k + 1)$ is true if $S(m)$ is true for all $1 \leq m \leq k$. By applying the induction step to $S(1), S(2), \cdots , S(\ell)$, we can conclude that $S(\ell + 1)$ is true. Then, we can also conclude that $S(\ell + 2)$ is true because $S(1), S(2), \cdots , S(\ell + 1)$ are true, $S(\ell + 3)$ is true because $S(1), S(2), \cdots , S(\ell + 2)$ are true, and so on. Thus, this method can prove that $S(n)$ is true for all $n \in \mathbb{N}$.

Similarly to weak induction, strong induction can also be modified slightly to prove a statement $S(n)$ for all $n \geq a$ for some natural number $a > 1$.

Strong Induction (Modified)

To prove a statement $S(n)$ for all $n \geq a$, we prove the following.

1. $S(a), S(a + 1), \cdots , S(a + \ell)$ are true for some natural number ℓ;

2. If $S(m)$ is true for all $a \leq m \leq k$ for some $k \geq a + \ell$, then $S(k + 1)$ is also true.

Since we are proving a statement $S(n)$ is true for all $n \geq a$, the base cases would start with $S(a)$ instead of $S(1)$. So, the base cases would be $S(a), S(a+1), \cdots , S(a+$

ℓ). Then, applying the induction step, we can conclude that $S(a+\ell+1)$, $S(a+\ell+2)$, $S(a+\ell+3)$, etc. are true. Thus, we can prove that $S(n)$ is true for all $n \geq a$.

Now let us discuss the difference between weak induction and strong induction. Note that in weak induction, we only need one previous case $S(k)$ to prove $S(k+1)$. On the other hand, in strong induction, we need to use at least one previous case other than $S(k)$ to prove $S(k+1)$. Usually, we use two or more previous cases in strong induction. However, like in problem 16, we do have some problems where we only use one previous case in strong induction. In problem 16, we use $S(k-1)$, NOT $S(k)$, to prove $S(k+1)$, so it is strong induction instead of weak induction. Thus, the difference between weak induction and strong induction mainly lies in the induction step. That is why you will not be sure about whether to use weak or strong induction and how many base cases you need to prove until you prove the induction step.

When we prove the induction step, we need to assume that the previous case(s) is/are true and then prove that $S(k+1)$ is true by using that assumption. The induction step could be quite challenging because sometimes you would need to be creative in connecting the previous cases to $S(k+1)$.

To be more familiar with mathematical induction and to be better at deciding between weak and strong inductions, the best way is to practice as much as possible, and this book provides plenty of such practice problems. For each of the problems in this book, I suggest that you try it on your own first and then look at the solution after you figure it out or get stuck for a long time.

Problem 1

Topic: Series

<table>
<tr><td>

Problem: Prove

$$\sum_{j=1}^{n} j = \frac{n(n+1)}{2}.$$

</td></tr>
</table>

Solution

Base Cases

We need to prove that it is true for $n = 1$. We can check that

$$\sum_{j=1}^{1} j = 1 = \frac{1(1+1)}{2},$$

so it is true for $n = 1$.

Induction Step

Assume

$$\sum_{j=1}^{k} j = \frac{k(k+1)}{2} \tag{1.1}$$

for some $k \geq 1$. To prove the induction step, we need to prove

$$\sum_{j=1}^{k+1} j = \frac{(k+1)(k+1+1)}{2} = \frac{(k+1)(k+2)}{2}.$$

Using assumption (1.1),

$$\sum_{j=1}^{k+1} j = \sum_{j=1}^{k} j + (k+1)$$

$$= \frac{k(k+1)}{2} + (k+1)$$

$$= \frac{k(k+1)}{2} + \frac{2(k+1)}{2}$$

$$= \frac{(k+1)(k+2)}{2}.$$

Thus, we proved that it is true for $n = k+1$ if it is true for $n = k$.

Conclusion

We proved that it is true for $n = 1$. Then, we proved that it is true for $n = k+1$ if it is true for $n = k$. Therefore, by weak induction,

$$\sum_{j=1}^{n} j = \frac{n(n+1)}{2}$$

for all $n \in \mathbb{N}$.

Problem 2

Topic: Series

Problem: Prove
$$\sum_{j=1}^{n} j^2 = \frac{n(n+1)(2n+1)}{6}.$$

Solution

Base Cases

We need to prove that it is true for $n = 1$. We can check that

$$\sum_{j=1}^{1} j^2 = 1^2 = 1 = \frac{1(1+1)(2\cdot 1+1)}{6},$$

so it is true for $n = 1$.

Induction Step

Assume

$$\sum_{j=1}^{k} j^2 = \frac{k(k+1)(2k+1)}{6} \tag{2.1}$$

for some $k \geq 1$. To prove the induction step, we need to prove

$$\sum_{j=1}^{k+1} j^2 = \frac{(k+1)(k+1+1)(2(k+1)+1)}{6} = \frac{(k+1)(k+2)(2k+3)}{6}.$$

Using assumption (2.1),

$$\sum_{j=1}^{k+1} j^2 = \sum_{j=1}^{k} j^2 + (k+1)^2$$

$$= \frac{k(k+1)(2k+1)}{6} + (k+1)^2$$

$$= \frac{k(k+1)(2k+1)}{6} + \frac{(k+1)(6k+6)}{6}$$

$$= \frac{(k+1)\left[k(2k+1) + (6k+6)\right]}{6}$$

$$= \frac{(k+1)\left(2k^2 + 7k + 6\right)}{6}$$

$$= \frac{(k+1)(k+2)(2k+3)}{6}.$$

Thus, we proved that it is true for $n = k+1$ if it is true for $n = k$.

Conclusion

We proved that it is true for $n = 1$. Then, we proved that it is true for $n = k+1$ if it is true for $n = k$. Therefore, by weak induction,

$$\sum_{j=1}^{n} j^2 = \frac{n(n+1)(2n+1)}{6}$$

for all $n \in \mathbb{N}$.

Problem 3

Topic: Series

Problem: Prove
$$\sum_{j=1}^{n} j^3 = \left[\frac{n(n+1)}{2}\right]^2.$$

Solution

Base Cases

We need to prove that it is true for $n = 1$. We can check that

$$\sum_{j=1}^{1} j^3 = 1^3 = 1 = \left[\frac{1(1+1)}{2}\right]^2,$$

so it is true for $n = 1$.

Induction Step

Assume

$$\sum_{j=1}^{k} j^3 = \left[\frac{k(k+1)}{2}\right]^2 \tag{3.1}$$

for some $k \geq 1$. To prove the induction step, we need to prove

$$\sum_{j=1}^{k+1} j^3 = \left[\frac{(k+1)(k+1+1)}{2}\right]^2 = \left[\frac{(k+1)(k+2)}{2}\right]^2.$$

Using assumption (3.1),

$$\sum_{j=1}^{k+1} j^3 = \sum_{j=1}^{k} j^3 + (k+1)^3$$

$$= \left[\frac{k(k+1)}{2}\right]^2 + (k+1)^3$$

$$= \frac{k^2(k+1)^2}{4} + \frac{(4k+4)(k+1)^2}{4}$$

$$= \frac{(k+1)^2\left(k^2+4k+4\right)}{4}$$

$$= \frac{(k+1)^2\left(k+2\right)^2}{4}$$

$$= \left[\frac{(k+1)(k+2)}{2}\right]^2.$$

Thus, we proved that it is true for $n = k+1$ if it is true for $n = k$.

Conclusion

We proved that it is true for $n = 1$. Then, we proved that it is true for $n = k+1$ if it is true for $n = k$. Therefore, by weak induction,

$$\sum_{j=1}^{n} j^3 = \left[\frac{n(n+1)}{2}\right]^2$$

for all $n \in \mathbb{N}$.

Problem 4

Topic: Harmonic Number, Inequality

Problem: Prove that $H_n \leq n$ for all $n \in \mathbb{N}$ where H_n denotes the n-th Harmonic number.

Definition - Harmonic Number:

$$H_n = \sum_{j=1}^{n} \frac{1}{j}.$$

Solution

Base Cases

We need to prove that it is true for $n = 1$. We can check that

$$H_1 = \sum_{j=1}^{1} \frac{1}{j} = \frac{1}{1} = 1 \leq 1,$$

so it is true for $n = 1$.

Induction Step

Assume

$$H_k \leq k \tag{4.1}$$

for some $k \geq 1$. To prove the induction step, we need to prove

$$H_{k+1} \leq k + 1.$$

Using assumption (4.1),

$$H_{k+1} = \sum_{j=1}^{k+1} \frac{1}{j}$$
$$= \sum_{j=1}^{k} \frac{1}{j} + \frac{1}{k+1}$$
$$= H_k + \frac{1}{k+1}$$
$$\leq k + \frac{1}{k+1}.$$

Since $k + 1 \geq 1$,

$$\frac{1}{k+1} \leq 1,$$

and we obtain

$$H_{k+1} \leq k + 1.$$

Thus, we proved that it is true for $n = k + 1$ if it is true for $n = k$.

Conclusion

We proved that it is true for $n = 1$. Then, we proved that it is true for $n = k + 1$ if it is true for $n = k$. Therefore, by weak induction,

$$H_n \leq n$$

for all $n \in \mathbb{N}$.

Problem 5

Topic: Harmonic Number, Inequality

Problem: Prove
$$H_n \geq \frac{2n}{n+1}$$
for all $n \in \mathbb{N}$ where H_n denotes the n-th Harmonic number.

Definition - Harmonic Number:
$$H_n = \sum_{j=1}^{n} \frac{1}{j}.$$

Solution

Base Cases

We need to prove that it is true for $n = 1$. We can check that

$$H_1 = \sum_{j=1}^{1} \frac{1}{j} = \frac{1}{1} = 1 \geq \frac{2 \cdot 1}{1 + 1},$$

so it is true for $n = 1$.

Induction Step

Assume

$$H_k \geq \frac{2k}{k+1} \tag{5.1}$$

for some $k \geq 1$. To prove the induction step, we need to prove

$$H_{k+1} \geq \frac{2(k+1)}{(k+1)+1} = \frac{2k+2}{k+2}.$$

Using assumption (5.1),

$$H_{k+1} = \sum_{j=1}^{k+1} \frac{1}{j}$$

$$= \sum_{j=1}^{k} \frac{1}{j} + \frac{1}{k+1}$$

$$= H_k + \frac{1}{k+1}$$

$$\geq \frac{2k}{k+1} + \frac{1}{k+1}$$

$$= \frac{2k+1}{k+1}.$$

Note that

$$\frac{2k+1}{k+1} - \frac{2k+2}{k+2} = \frac{k}{(k+1)(k+2)} \geq 0$$

because $k \geq 0$, $k+1 > 0$, and $k+2 > 0$, so

$$\frac{2k+1}{k+1} \geq \frac{2k+2}{k+2}.$$

Therefore,

$$H_{k+1} \geq \frac{2k+2}{k+2}.$$

Thus, we proved that it is true for $n = k+1$ if it is true for $n = k$.

Conclusion

We proved that it is true for $n = 1$. Then, we proved that it is true for $n = k+1$ if it is true for $n = k$. Therefore, by weak induction,

$$H_n \geq \frac{2n}{n+1}$$

for all $n \in \mathbb{N}$.

Problem 6

Topic: Inequality

Problem: Prove $e^n > n + 1$ for all $n \in \mathbb{N}$ where e is the base of natural logarithm, $e \approx 2.7183$.

Solution

Base Cases

We need to prove that it is true for $n = 1$. We can check that

$$e^1 = e \approx 2.7183 > 1 + 1,$$

so it is true for $n = 1$.

Induction Step

Assume

$$e^k > k + 1 \tag{6.1}$$

for some $k \geq 1$. To prove the induction step, we need to prove

$$e^{k+1} > (k + 1) + 1 = k + 2.$$

Multiplying e to both sides of the inequality in assumption (6.1),

$$e^{k+1} > ek + e.$$

Since $e > 2 > 1$,

$$ek + e > k + 2.$$

Therefore,

$$e^{k+1} > k + 2.$$

Thus, we proved that it is true for $n = k + 1$ if it is true for $n = k$.

Conclusion

We proved that it is true for $n = 1$. Then, we proved that it is true for $n = k + 1$ if it is true for $n = k$. Therefore, by weak induction,

$$e^n > n + 1$$

for all $n \in \mathbb{N}$.

Problem 7

Topic: Inequality

Problem: Prove $e^{n-1} > 2^n$ for $n \geq 4$ where e is the base of natural logarithm, $e \approx 2.7183$.

Solution

Base Cases

We need to prove that it is true for $n = 4$. We can check that

$$e^{4-1} = e^3 \approx (2.7183)^3 > (2.7)^3 = 19.683 > 2^4,$$

so it is true for $n = 4$.

Induction Step

Assume

$$e^{k-1} > 2^k \tag{7.1}$$

for some $k \geq 4$. To prove the induction step, we need to prove

$$e^{(k+1)-1} = e^k > 2^{k+1}.$$

Multiplying e to both sides of the inequality in assumption (7.1),

$$e^k > e \cdot 2^k.$$

Since $e > 2$,

$$e \cdot 2^k > 2 \cdot 2^k = 2^{k+1}.$$

Therefore,

$$e^k > 2^{k+1}.$$

Thus, we proved that it is true for $n = k + 1$ if it is true for $n = k$.

Conclusion

We proved that it is true for $n = 4$. Then, we proved that it is true for $n = k + 1$ if it is true for $n = k$. Therefore, by weak induction,

$$e^{n-1} > 2^n$$

for $n \geq 4$.

Problem 8

Topic: Inequality

Problem: Prove $n! > 2n + 2^n$ for $n \geq 5$.

Definition - Factorial:

$$n! = n \cdot (n-1) \cdot (n-2) \cdots 2 \cdot 1.$$

$$0! = 1.$$

Solution

Base Cases

We need to prove that it is true for $n = 5$. We can check that

$$5! = 120 > 42 = 2 \cdot 5 + 2^5,$$

so it is true for $n = 5$.

Induction Step

Assume

$$k! > 2k + 2^k \tag{8.1}$$

for some $k \geq 5$. To prove the induction step, we need to prove

$$(k+1)! > 2(k+1) + 2^{k+1}.$$

Multiplying both sides of assumption (8.1) by $k + 1$,

$$(k+1)! > 2k(k+1) + 2^k(k+1).$$

Since $k \geq 5 > 1$ and $k + 1 \geq 6 > 2$,

$$2k(k + 1) > 2(k + 1)$$

and

$$2^k(k + 1) > 2^k \cdot 2 = 2^{k+1}.$$

So,

$$(k + 1)! > 2(k + 1) + 2^{k+1}.$$

Thus, we proved that it is true for $n = k + 1$ if it is true for $n = k$.

Conclusion

We proved that it is true for $n = 5$. Then, we proved that it is true for $n = k + 1$ if it is true for $n = k$. Therefore, by weak induction,

$$n! > 2n + 2^n$$

for $n \geq 5$.

Problem 9

Topic: Inequality

Problem: Prove $n^n > 3n^2 + 6$ for $n \geq 4$.

Solution

Base Cases

We need to prove that it is true for $n = 4$. We can check that

$$4^4 > 3 \cdot 4^2 + 6,$$

so it is true for $n = 4$.

Induction Step

Assume
$$k^k > 3k^2 + 6 \tag{9.1}$$

for some $k \geq 4$. To prove the induction step, we need to prove

$$(k+1)^{k+1} > 3(k+1)^2 + 6 = 3k^2 + 6k + 9.$$

Since $k + 1 > k$,

$$(k+1)^{k+1} = (k+1) \cdot (k+1)^k > (k+1) \cdot k^k.$$

Using assumption (9.1),

$$(k+1) \cdot k^k > (k+1)\left(3k^2 + 6\right) = 3k^3 + 3k^2 + 6k + 6.$$

Since $k \geq 4$,

$$3k^3 + 6 \geq 3 \cdot 4^3 + 6 > 9,$$

so

$$3k^3 + 3k^2 + 6k + 6 > 3k^2 + 6k + 9.$$

Therefore,

$$(k+1)^{k+1} > 3k^2 + 6k + 9 = 3(k+1)^2 + 6.$$

Thus, we proved that it is true for $n = k + 1$ if it is true for $n = k$.

Conclusion

We proved that it is true for $n = 4$. Then, we proved that it is true for $n = k + 1$ if it is true for $n = k$. Therefore, by weak induction,

$$n^n > 3n^2 + 6$$

for $n \geq 4$.

Problem 10

Topic: Inequality

> **Problem:** Prove Bernoulli's inequality, which states that
> $$(1+x)^n \geq 1 + nx$$
> for $x > -1$ and $n \in \mathbb{N}$.

Solution

Base Cases

We need to prove that it is true for $n = 1$. Since

$$(1+x)^1 = 1 + x$$

and

$$1 + 1 \cdot x = 1 + x,$$

it is indeed true that

$$(1+x)^1 \geq 1 + 1 \cdot x.$$

Induction Step

Assume

$$(1+x)^k \geq 1 + kx \tag{10.1}$$

for some $k \geq 1$. To prove the induction step, we need to prove

$$(1+x)^{k+1} \geq 1 + (k+1)x.$$

Since $x > -1$, $1 + x > 0$. So, we can multiply $1 + x$ to both sides of the inequality in assumption (10.1) without changing the sign:

$$(1 + x)^{k+1} \geq (1 + x)(1 + kx)$$
$$= 1 + (k + 1)x + kx^2.$$

Since $k > 0$ and $x^2 \geq 0$, $kx^2 \geq 0$, and thus

$$(1 + x)^{k+1} \geq 1 + (k + 1)x.$$

Thus, we proved that it is true for $n = k + 1$ if it is true for $n = k$.

Conclusion

We proved that it is true for $n = 1$. Then, we proved that it is true for $n = k + 1$ if it is true for $n = k$. Therefore, by weak induction,

$$(1 + x)^n \geq 1 + nx$$

if $x > -1$ for all $n \in \mathbb{N}$.

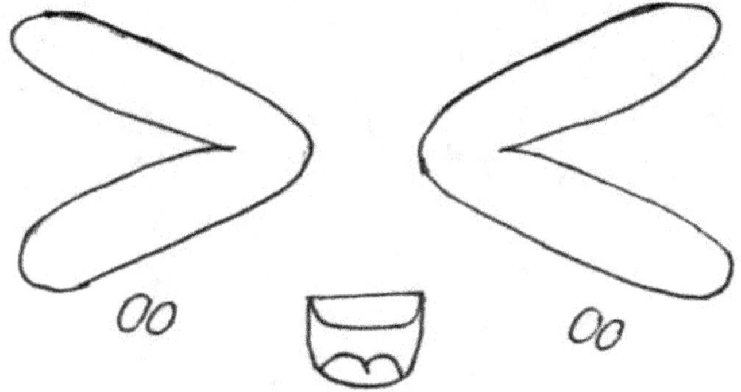

Problem 11

Topic: Series

> **Problem:** For $a \neq 1$, prove
> $$\sum_{j=0}^{n-1} a^j = \frac{a^n - 1}{a - 1}.$$

Solution

Base Cases

We need to prove that it is true for $n = 1$. We can check that

$$\sum_{j=0}^{1-1} a^j = a^0 = 1 = \frac{a^1 - 1}{a - 1},$$

so it is true for $n = 1$.

Induction Step

Assume

$$\sum_{j=0}^{k-1} a^j = \frac{a^k - 1}{a - 1} \tag{11.1}$$

for some $k \geq 1$. To prove the induction step, we need to prove

$$\sum_{j=0}^{(k+1)-1} a^j = \sum_{j=0}^{k} a^j = \frac{a^{k+1} - 1}{a - 1}.$$

Using assumption (11.1),

$$\sum_{j=0}^{k} a^j = \sum_{j=0}^{k-1} a^j + a^k$$

$$= \frac{a^k - 1}{a - 1} + a^k$$

$$= \frac{a^k - 1}{a - 1} + \frac{a^k(a - 1)}{a - 1}$$

$$= \frac{a^k - 1}{a - 1} + \frac{a^{k+1} - a^k}{a - 1}$$

$$= \frac{a^{k+1} - 1}{a - 1}.$$

Thus, we proved that it is true for $n = k + 1$ if it is true for $n = k$.

Conclusion

We proved that it is true for $n = 1$. Then, we proved that it is true for $n = k + 1$ if it is true for $n = k$. Therefore, by weak induction,

$$\sum_{j=0}^{n-1} a^j = \frac{a^n - 1}{a - 1}$$

if $a \neq 1$ for all $n \in \mathbb{N}$.

Problem 12

Topic: Series, Fibonacci Number

Problem: Prove
$$\sum_{j=1}^{n} F_j = F_{n+2} - 1$$
where F_n denotes the n-th Fibonacci number.

Definition - Fibonacci Numbers: The Fibonacci numbers are a sequence of numbers defined as $F_0 = 0$, $F_1 = 1$, and $F_n = F_{n-1} + F_{n-2}$ for $n \geq 2$.

Solution

Base Cases

We need to prove that it is true for $n = 1$. When $n = 1$,

$$\sum_{j=1}^{1} F_j = F_1 = 1$$

and

$$F_{1+2} - 1 = F_3 - 1 = F_2 + F_1 - 1 = F_1 + F_0 + F_1 - 1 = 1 + 0 + 1 - 1 = 1,$$

so it is indeed true that

$$\sum_{j=1}^{1} F_j = F_{1+2} - 1.$$

Induction Step

Assume

$$\sum_{j=1}^{k} F_j = F_{k+2} - 1 \tag{12.1}$$

for some $k \geq 1$. To prove the induction step, we need to prove

$$\sum_{j=1}^{k+1} F_j = F_{(k+1)+2} - 1 = F_{k+3} - 1.$$

Using assumption (12.1),

$$\sum_{j=1}^{k+1} F_j = \sum_{j=1}^{k} F_j + F_{k+1}$$
$$= F_{k+2} - 1 + F_{k+1}.$$

Using the recursive relation of F_n,

$$F_n = F_{n-1} + F_{n-2},$$

we have

$$F_{k+2} + F_{k+1} = F_{k+3}.$$

Therefore,

$$\sum_{j=1}^{k+1} F_j = F_{k+3} - 1.$$

Thus, we proved that it is true for $n = k + 1$ if it is true for $n = k$.

Conclusion

We proved that it is true for $n = 1$. Then, we proved that it is true for $n = k + 1$ if it is true for $n = k$. Therefore, by weak induction,

$$\sum_{j=1}^{n} F_j = F_{n+2} - 1$$

for all $n \in \mathbb{N}$.

Problem 13

Topic: Series, Fibonacci Number

Problem: Prove
$$\sum_{j=0}^{n} F_{2j+1} = F_{2n+2}$$
where F_n denotes the n-th Fibonacci number.

Definition - Fibonacci Numbers: The Fibonacci numbers are a sequence of numbers defined as $F_0 = 0$, $F_1 = 1$, and $F_n = F_{n-1} + F_{n-2}$ for $n \geq 2$.

Solution

Base Cases

We need to prove that it is true for $n = 1$. When $n = 1$,

$$\sum_{j=0}^{1} F_{2j+1} = F_1 + F_3$$

and

$$F_{2 \cdot 1 + 2} = F_4 = F_2 + F_3.$$

Since

$$F_2 = F_1 + F_0 = F_1 + 0 = F_1,$$

it is indeed true that

$$\sum_{j=0}^{1} F_{2j+1} = F_{2 \cdot 1 + 2}.$$

Induction Step

Assume

$$\sum_{j=0}^{k} F_{2j+1} = F_{2k+2} \tag{13.1}$$

for some $k \geq 1$. To prove the induction step, we need to prove

$$\sum_{j=0}^{k+1} F_{2j+1} = F_{2(k+1)+2} = F_{2k+4}.$$

Using assumption (13.1),

$$\sum_{j=0}^{k+1} F_{2j+1} = \sum_{j=0}^{k} F_{2j+1} + F_{2k+3}$$
$$= F_{2k+2} + F_{2k+3}.$$

Using the recursive relation of F_n,

$$F_n = F_{n-1} + F_{n-2},$$

we have

$$F_{2k+2} + F_{2k+3} = F_{2k+4}.$$

Therefore,

$$\sum_{j=0}^{k+1} F_{2j+1} = F_{2k+4}.$$

Thus, we proved that it is true for $n = k + 1$ if it is true for $n = k$.

Conclusion

We proved that it is true for $n = 1$. Then, we proved that it is true for $n = k + 1$ if it is true for $n = k$. Therefore, by weak induction,

$$\sum_{j=0}^{n} F_{2j+1} = F_{2n+2}$$

for all $n \in \mathbb{N}$.

Problem 14

Topic: Series, Fibonacci Number

Problem: Prove
$$\sum_{j=1}^{n} F_{2j} = F_{2n+1} - 1$$
where F_n denotes the n-th Fibonacci number.

Definition - Fibonacci Numbers: The Fibonacci numbers are a sequence of numbers defined as $F_0 = 0$, $F_1 = 1$, and $F_n = F_{n-1} + F_{n-2}$ for $n \geq 2$.

Solution

Base Cases

We need to prove that it is true for $n = 1$. When $n = 1$,

$$\sum_{j=1}^{1} F_{2j} = F_2 = F_1 + F_0 = 1 + 0 = 1$$

and

$$F_{2\cdot 1+1} - 1 = F_3 - 1 = F_2 + F_1 - 1 = F_1 + F_0 + F_1 - 1 = 1 + 0 + 1 - 1 = 1,$$

so it is indeed true that

$$\sum_{j=1}^{1} F_{2j} = F_{2\cdot 1+1} - 1.$$

Induction Step

Assume

$$\sum_{j=1}^{k} F_{2j} = F_{2k+1} - 1 \tag{14.1}$$

for some $k \geq 1$. To prove the induction step, we need to prove

$$\sum_{j=1}^{k+1} F_{2j} = F_{2(k+1)+1} - 1 = F_{2k+3} - 1.$$

Using assumption (14.1),

$$\sum_{j=1}^{k+1} F_{2j} = \sum_{j=1}^{k} F_{2j} + F_{2k+2}$$

$$= F_{2k+1} - 1 + F_{2k+2}.$$

Using the recursive relation of F_n,

$$F_n = F_{n-1} + F_{n-2},$$

we have

$$F_{2k+1} + F_{2k+2} = F_{2k+3}.$$

Therefore,

$$\sum_{j=1}^{k+1} F_{2j} = F_{2k+3} - 1.$$

Thus, we proved that it is true for $n = k + 1$ if it is true for $n = k$.

Conclusion

We proved that it is true for $n = 1$. Then, we proved that it is true for $n = k + 1$ if it is true for $n = k$. Therefore, by weak induction,

$$\sum_{j=1}^{n} F_{2j} = F_{2n+1} - 1$$

for all $n \in \mathbb{N}$.

Problem 15

Topic: Series, Fibonacci Number

Problem: Prove

$$\sum_{j=1}^{n} F_j^2 = F_n F_{n+1}$$

where F_n denotes the n-th Fibonacci number.

Definition - Fibonacci Numbers: The Fibonacci numbers are a sequence of numbers defined as $F_0 = 0$, $F_1 = 1$, and $F_n = F_{n-1} + F_{n-2}$ for $n \geq 2$.

Solution

Base Cases

We need to prove that it is true for $n = 1$. When $n = 1$,

$$\sum_{j=1}^{1} F_j^2 = F_1^2 = 1$$

and

$$F_1 F_{1+1} = F_1 F_2 = F_1(F_1 + F_0) = 1,$$

so it is indeed true that

$$\sum_{j=1}^{1} F_j^2 = F_1 F_{1+1}.$$

Induction Step

Assume

$$\sum_{j=1}^{k} F_j^2 = F_k F_{k+1} \tag{15.1}$$

for some $k \geq 1$. To prove the induction step, we need to prove

$$\sum_{j=1}^{k+1} F_j^2 = F_{k+1} F_{(k+1)+1} = F_{k+1} F_{k+2}.$$

Using assumption (15.1),

$$\begin{aligned}
\sum_{j=1}^{k+1} F_j^2 &= \sum_{j=1}^{k} F_j^2 + F_{k+1}^2 \\
&= F_k F_{k+1} + F_{k+1}^2 \\
&= F_{k+1}(F_k + F_{k+1}).
\end{aligned}$$

Using the recursive relation of F_n,

$$F_n = F_{n-1} + F_{n-2},$$

we have

$$F_k + F_{k+1} = F_{k+2}.$$

Therefore,

$$\sum_{j=1}^{k+1} F_j^2 = F_{k+1} F_{k+2}.$$

Thus, we proved that it is true for $n = k + 1$ if it is true for $n = k$.

Conclusion

We proved that it is true for $n = 1$. Then, we proved that it is true for $n = k + 1$ if it is true for $n = k$. Therefore, by weak induction,

$$\sum_{j=1}^{n} F_j^2 = F_n F_{n+1}$$

for all $n \in \mathbb{N}$.

Problem 16

Topic: Series

Problem: Prove
$$\sum_{j=0}^{n} \frac{1}{2^n + 4^j} = \frac{n+1}{2^{n+1}}.$$

Solution

Base Cases

We need to prove that it is true for $n = 1$ and $n = 2$. We can check that

$$\sum_{j=0}^{1} \frac{1}{2^1 + 4^j} = \frac{1}{2+1} + \frac{1}{2+4} = \frac{1}{2} = \frac{1+1}{2^{1+1}}$$

and

$$\sum_{j=0}^{2} \frac{1}{2^2 + 4^j} = \frac{1}{4+1} + \frac{1}{4+4} + \frac{1}{4+16} = \frac{3}{8} = \frac{2+1}{2^{2+1}},$$

so it is true for $n = 1$ and $n = 2$.

Induction Step

Assume

$$\sum_{j=0}^{m} \frac{1}{2^m + 4^j} = \frac{m+1}{2^{m+1}} \tag{16.1}$$

for all $1 \le m \le k$ for some $k \ge 2$. To prove the induction step, we need to prove

$$\sum_{j=0}^{k+1} \frac{1}{2^{k+1} + 4^j} = \frac{(k+1)+1}{2^{(k+1)+1}} = \frac{k+2}{2^{k+2}}.$$

Note that we can write the sum when $n = k + 1$ as

$$\sum_{j=0}^{k+1} \frac{1}{2^{k+1} + 4^j} = \frac{1}{2^{k+1} + 1} + \frac{1}{2^{k+1} + 4^{k+1}} + \sum_{j=1}^{k} \frac{1}{2^{k+1} + 4^j}$$

$$= \frac{1}{2^{k+1} + 1} + \frac{1}{2^{k+1} + 4^{k+1}} + \frac{1}{4} \sum_{j=1}^{k} \frac{1}{2^{k-1} + 4^{j-1}}.$$

Substituting $j - 1 \to j$,

$$\sum_{j=1}^{k} \frac{1}{2^{k-1} + 4^{j-1}} = \sum_{j=0}^{k-1} \frac{1}{2^{k-1} + 4^j}.$$

Then, by assumption (16.1) when $m = k - 1$,

$$\sum_{j=0}^{k-1} \frac{1}{2^{k-1} + 4^j} = \frac{(k-1) + 1}{2^{(k-1)+1}} = \frac{k}{2^k}.$$

So,

$$\sum_{j=0}^{k+1} \frac{1}{2^{k+1} + 4^j} = \frac{1}{2^{k+1} + 1} + \frac{1}{2^{k+1} + 4^{k+1}} + \frac{1}{4} \cdot \frac{k}{2^k}$$

$$= \frac{1}{2^{k+1} + 1} + \frac{1}{2^{k+1} \left(1 + 2^{k+1}\right)} + \frac{k}{2^{k+2}}$$

$$= \frac{2^{k+1} + 1}{2^{k+1} \left(2^{k+1} + 1\right)} + \frac{k}{2^{k+2}}$$

$$= \frac{1}{2^{k+1}} + \frac{k}{2^{k+2}}$$

$$= \frac{k + 2}{2^{k+2}}.$$

Thus, we proved that it is true for $n = k+1$ if it is true for $n = m$ for all $1 \le m \le k$.

Conclusion

We proved that it is true for $n = 1$ and $n = 2$. Then, we proved that it is true for $n = k + 1$ if it is true for $n = m$ for all $1 \le m \le k$. Therefore, by strong induction,

$$\sum_{j=0}^{n} \frac{1}{2^n + 4^j} = \frac{n + 1}{2^{n+1}}$$

for all $n \in \mathbb{N}$.

Problem 17

Topic: Divisibility

Problem: Prove
$$3 \mid \left(n^3 - n\right)$$
for all $n \in \mathbb{N}$.

Definition - Divisibility: For $a \in \mathbb{Z} \setminus \{0\}$ and $b \in \mathbb{Z}$, a divides b, denoted as $a \mid b$, when a is a factor of b. Formally,

$$a \mid b \Leftrightarrow b = a\ell$$

for some $\ell \in \mathbb{Z}$.

Solution

Base Cases

We need to prove that it is true for $n = 1$. Since

$$1^3 - 1 = 0 = 3 \cdot 0$$

where 0 is an integer,

$$3 \mid \left(1^3 - 1\right)$$

by definition of divisibility.

Induction Step

Assume

$$3 \mid \left(k^3 - k\right) \tag{17.1}$$

for some $k \geq 1$. To prove the induction step, we need to prove

$$3 \mid \left((k+1)^3 - (k+1)\right).$$

By assumption (17.1) and the definition of divisibility,

$$k^3 - k = 3\ell$$

for some $\ell \in \mathbb{Z}$. Then,

$$\begin{aligned}
(k+1)^3 - (k+1) &= k^3 + 3k^2 + 3k + 1 - k - 1 \\
&= 3\ell + 3k^2 + 3k \\
&= 3\left(\ell + k^2 + k\right).
\end{aligned}$$

Since $\ell + k^2 + k$ is an integer,

$$3 \mid \left((k+1)^3 - (k+1)\right)$$

by definition of divisibility. Thus, we proved that it is true for $n = k+1$ if it is true for $n = k$.

Conclusion

We proved that it is true for $n = 1$. Then, we proved that it is true for $n = k+1$ if it is true for $n = k$. Therefore, by weak induction,

$$3 \mid \left(n^3 - n\right)$$

for all $n \in \mathbb{N}$.

Problem 18

Topic: Divisibility

Problem: Prove
$$5 \mid \left(n^5 - n\right)$$
for all $n \in \mathbb{N}$.

Definition - Divisibility: For $a \in \mathbb{Z} \setminus \{0\}$ and $b \in \mathbb{Z}$, a divides b, denoted as $a \mid b$, when a is a factor of b. Formally,
$$a \mid b \Leftrightarrow b = a\ell$$
for some $\ell \in \mathbb{Z}$.

Solution

Base Cases

We need to prove that it is true for $n = 1$. Since
$$1^5 - 1 = 0 = 5 \cdot 0$$
where 0 is an integer,
$$5 \mid \left(1^5 - 1\right)$$
by definition of divisibility.

Induction Step

Assume
$$5 \mid \left(k^5 - k\right) \tag{18.1}$$

for some $k \geq 1$. To prove the induction step, we need to prove

$$5 \mid \left((k+1)^5 - (k+1)\right).$$

By assumption (18.1) and the definition of divisibility,

$$k^5 - k = 5\ell$$

for some $\ell \in \mathbb{Z}$. Then,

$$\begin{aligned}
(k+1)^5 - (k+1) &= k^5 + 5k^4 + 10k^3 + 10k^2 + 5k + 1 - k - 1 \\
&= 5\ell + 5k^4 + 10k^3 + 10k^2 + 5k \\
&= 5\left(\ell + k^4 + 2k^3 + 2k^2 + k\right).
\end{aligned}$$

Since $\ell + k^4 + 2k^3 + 2k^2 + k$ is an integer,

$$5 \mid \left((k+1)^5 - (k+1)\right)$$

by definition of divisibility. Thus, we proved that it is true for $n = k+1$ if it is true for $n = k$.

Conclusion

We proved that it is true for $n = 1$. Then, we proved that it is true for $n = k+1$ if it is true for $n = k$. Therefore, by weak induction,

$$5 \mid \left(n^5 - n\right)$$

for all $n \in \mathbb{N}$.

Problem 19

Topic: Divisibility

Problem: Prove
$$7 \mid (9^n - 2^n)$$
for all $n \in \mathbb{N}$.

Definition - Divisibility: For $a \in \mathbb{Z} \setminus \{0\}$ and $b \in \mathbb{Z}$, a divides b, denoted as $a \mid b$, when a is a factor of b. Formally,
$$a \mid b \Leftrightarrow b = a\ell$$
for some $\ell \in \mathbb{Z}$.

Solution

Base Cases

We need to prove that it is true for $n = 1$. Since
$$9^1 - 2^1 = 7 = 7 \cdot 1$$
where 1 is an integer,
$$7 \mid (9^1 - 2^1)$$
by definition of divisibility.

Induction Step

Assume
$$7 \mid (9^k - 2^k) \tag{19.1}$$

for some $k \geq 1$. To prove the induction step, we need to prove

$$7 \mid \left(9^{k+1} - 2^{k+1}\right).$$

By assumption (19.1) and by definition of divisibility,

$$9^k - 2^k = 7\ell$$

for some $\ell \in \mathbb{Z}$. Then,

$$\begin{aligned}
9^{k+1} - 2^{k+1} &= 9 \cdot 9^k - 2 \cdot 2^k \\
&= (7 + 2) \cdot 9^k - 2 \cdot 2^k \\
&= 7 \cdot 9^k + 2 \cdot \left(9^k - 2^k\right) \\
&= 7 \cdot 9^k + 2 \cdot 7\ell \\
&= 7 \left(9^k + 2\ell\right).
\end{aligned}$$

Since $9^k + 2\ell$ is an integer,

$$7 \mid \left(9^{k+1} - 2^{k+1}\right)$$

by definition of divisibility. Thus, we proved that it is true for $n = k + 1$ if it is true for $n = k$.

Conclusion

We proved that it is true for $n = 1$. Then, we proved that it is true for $n = k + 1$ if it is true for $n = k$. Therefore, by weak induction,

$$7 \mid \left(9^n - 2^n\right)$$

for all $n \in \mathbb{N}$.

Problem 20

Topic: Divisibility

Problem: Prove
$$9 \mid \left(8^{2n} - 3 \cdot 7^n + 2\right)$$
for all $n \in \mathbb{N}$.

Definition - Divisibility: For $a \in \mathbb{Z} \setminus \{0\}$ and $b \in \mathbb{Z}$, a divides b, denoted as $a \mid b$, when a is a factor of b. Formally,

$$a \mid b \Leftrightarrow b = a\ell$$

for some $\ell \in \mathbb{Z}$.

Solution

Base Cases

We need to prove that it is true for $n = 1$. Since

$$8^{2 \cdot 1} - 3 \cdot 7^1 + 2 = 45 = 9 \cdot 5$$

where 5 is an integer,

$$9 \mid \left(8^{2 \cdot 1} - 3 \cdot 7^1 + 2\right)$$

by definition of divisibility.

Induction Step

Assume

$$9 \mid \left(8^{2k} - 3 \cdot 7^k + 2\right) \tag{20.1}$$

for some $k \geq 1$. To prove the induction step, we need to prove

$$9 \mid \left(8^{2(k+1)} - 3 \cdot 7^{k+1} + 2\right).$$

By assumption (20.1) and by definition of divisibility,

$$8^{2k} - 3 \cdot 7^k + 2 = 9\ell$$

for some $\ell \in \mathbb{Z}$. Then,

$$
\begin{aligned}
8^{2(k+1)} - 3 \cdot 7^{k+1} + 2 &= 64 \cdot 8^{2k} - 7 \cdot 3 \cdot 7^k + 2 \\
&= 63 \cdot 8^{2k} - 6 \cdot 3 \cdot 7^k + 8^{2k} - 3 \cdot 7^k + 2 \\
&= 63 \cdot 8^{2k} - 6 \cdot 3 \cdot 7^k + 9\ell \\
&= 9\left(7 \cdot 8^{2k} - 2 \cdot 7^k + \ell\right).
\end{aligned}
$$

Since $7 \cdot 8^{2k} - 2 \cdot 7^k + \ell$ is an integer,

$$9 \mid \left(8^{2(k+1)} - 3 \cdot 7^{k+1} + 2\right)$$

by definition of divisibility. Thus, we proved that it is true for $n = k + 1$ if it is true for $n = k$.

Conclusion

We proved that it is true for $n = 1$. Then, we proved that it is true for $n = k + 1$ if it is true for $n = k$. Therefore, by weak induction,

$$9 \mid \left(8^{2n} - 3 \cdot 7^n + 2\right)$$

for all $n \in \mathbb{N}$.

Problem 21

Topic: Inequality

Problem: For $x, y \geq 0$, prove

$$\left(\frac{x+y}{2}\right)^n \leq \frac{x^n + y^n}{2}$$

for all $n \in \mathbb{N}$.

Solution

Base Cases

We need to prove that it is true for $n = 1$. We can check that

$$\left(\frac{x+y}{2}\right)^1 = \frac{x+y}{2} \leq \frac{x^1 + y^1}{2},$$

so it is true for $n = 1$.

Induction Step

Assume

$$\left(\frac{x+y}{2}\right)^k \leq \frac{x^k + y^k}{2} \tag{21.1}$$

for some $k \geq 1$. To prove the induction step, we need to prove

$$\left(\frac{x+y}{2}\right)^{k+1} \leq \frac{x^{k+1} + y^{k+1}}{2}.$$

Multiplying both sides of the inequality in assumption (21.1) by $\frac{x+y}{2}$,

$$\left(\frac{x+y}{2}\right)^{k+1} \leq \left(\frac{x+y}{2}\right)\left(\frac{x^k+y^k}{2}\right)$$

$$= \frac{x^{k+1}+y^{k+1}+xy^k+yx^k}{4}$$

$$= \frac{x^{k+1}+y^{k+1}}{2} - \frac{x^{k+1}+y^{k+1}}{4} + \frac{xy^k+yx^k}{4}$$

$$= \frac{x^{k+1}+y^{k+1}}{2} - \frac{x^{k+1}+y^{k+1}-xy^k-yx^k}{4}$$

$$= \frac{x^{k+1}+y^{k+1}}{2} - \frac{\left(x^k-y^k\right)(x-y)}{4}.$$

For any real numbers x and y, it is always either $x \geq y$ or $y \geq x$. If $x \geq y$, then $x - y \geq 0$ and $x^k - y^k \geq 0$, so

$$\left(x^k-y^k\right)(x-y) \geq 0.$$

If $y \geq x$, then $x - y \leq 0$ and $x^k - y^k \leq 0$, so

$$\left(x^k-y^k\right)(x-y) \geq 0.$$

In both cases,

$$\left(x^k-y^k\right)(x-y) \geq 0.$$

So,

$$-\frac{\left(x^k-y^k\right)(x-y)}{4} \leq 0,$$

and thus we have

$$\frac{x^{k+1}+y^{k+1}}{2} - \frac{\left(x^k-y^k\right)(x-y)}{4} \leq \frac{x^{k+1}+y^{k+1}}{2}.$$

Therefore,

$$\left(\frac{x+y}{2}\right)^{k+1} \leq \frac{x^{k+1}+y^{k+1}}{2}.$$

Thus, we proved that it is true for $n = k+1$ if it is true for $n = k$.

Conclusion

We proved that it is true for $n = 1$. Then, we proved that it is true for $n = k+1$ if it is true for $n = k$. Therefore, by weak induction,

$$\left(\frac{x+y}{2}\right)^n \leq \frac{x^n+y^n}{2}$$

for $x, y \geq 0$ and for all $n \in \mathbb{N}$.

Problem 22

Topic: Inequality

Problem: If $a > b > 0$, prove

$$(2a - 2b)^n \geq 2a^n - (2b)^n$$

for all $n \in \mathbb{N}$.

Solution

Base Cases

We need to prove that it is true for $n = 1$. We can check that

$$(2a - 2b)^1 = 2a - 2b \geq 2a^1 - (2b)^1,$$

so it is true for $n = 1$.

Induction Step

Assume

$$(2a - 2b)^k \geq 2a^k - (2b)^k \tag{22.1}$$

for some $k \geq 1$. To prove the induction step, we need to prove

$$(2a - 2b)^{k+1} \geq 2a^{k+1} - (2b)^{k+1}.$$

Since $a > b$, $2a - 2b > 0$, so we can multiply $2a - 2b$ to both sides of the inequality in assumption (22.1) without changing the sign:

$$
\begin{aligned}
(2a - 2b)^{k+1} &\geq (2a - 2b)\left(2a^k - (2b)^k\right) \\
&= 4a^{k+1} + (2b)^{k+1} - 4ba^k - (2a)(2b)^k \\
&= 2a^{k+1} + 2a^{k+1} - (2b)^{k+1} + 2(2b)^{k+1} - 4ba^k - (2a)(2b)^k \\
&= 2a^{k+1} - (2b)^{k+1} + 2(a - 2b)\left(a^k - (2b)^k\right).
\end{aligned}
$$

For any real numbers a and $2b$, it is always either $a \geq 2b$ or $2b \geq a$. If $a \geq 2b$, then $a - 2b \geq 0$ and $a^k - (2b)^k \geq 0$, so

$$(a - 2b)\left(a^k - (2b)^k\right) \geq 0.$$

If $2b \geq a$, then $a - 2b \leq 0$ and $a^k - (2b)^k \leq 0$, so

$$(a - 2b)\left(a^k - (2b)^k\right) \geq 0.$$

In both cases,
$$(a - 2b)\left(a^k - (2b)^k\right) \geq 0.$$

So,
$$2a^{k+1} - (2b)^{k+1} + 2(a - 2b)\left(a^k - (2b)^k\right) \geq 2a^{k+1} - (2b)^{k+1}.$$

Therefore,
$$(2a - 2b)^{k+1} \geq 2a^{k+1} - (2b)^{k+1}.$$

Thus, we proved that it is true for $n = k + 1$ if it is true for $n = k$.

Conclusion

We proved that it is true for $n = 1$. Then, we proved that it is true for $n = k + 1$ if it is true for $n = k$. Therefore, by weak induction,

$$(2a - 2b)^n \geq 2a^n - (2b)^n$$

if $a > b > 0$ for all $n \in \mathbb{N}$.

Problem 23

Topic: Inequality

Problem: For $y \geq 0$, prove

$$ny - y^n \leq n - 1$$

for all $n \in \mathbb{N}$.

Solution

Base Cases

We need to prove that it is true for $n = 1$ and $n = 2$. We can check that

$$1 \cdot y - y^1 = 0 \leq 1 - 1$$

and

$$2 \cdot y - y^2 = -(y - 1)^2 + 1 \leq 1 = 2 - 1,$$

so it is true for $n = 1$ and $n = 2$.

Induction Step

Assume

$$my - y^m \leq m - 1 \tag{23.1}$$

for all $1 \leq m \leq k$ for some $k \geq 2$. To prove the induction step, we need to prove

$$(k + 1)y - y^{k+1} \leq (k + 1) - 1 = k.$$

By assumption (23.1) when $m = k$,

$$ky - y^k \leq k - 1 \Leftrightarrow -y^k \leq k - 1 - ky.$$

Since $y \geq 0$, we can multiply y to both sides of the inequality without changing the sign:

$$-y^{k+1} \leq y(k - 1 - ky).$$

So,

$$(k + 1)y - y^{k+1} \leq (k + 1)y + y(k - 1 - ky)$$
$$= 2ky - ky^2$$
$$= k\left(2y - y^2\right).$$

By assumption (23.1) when $m = 2$,

$$2y - y^2 \leq 2 - 1 = 1.$$

Therefore,

$$(k + 1)y - y^{k+1} \leq k.$$

Thus, we proved that it is true for $n = k+1$ if it is true for $n = m$ for all $1 \leq m \leq k$.

Conclusion

We proved that it is true for $n = 1$ and $n = 2$. Then, we proved that it is true for $n = k + 1$ if it is true for $n = m$ for all $1 \leq m \leq k$. Therefore, by strong induction,

$$ny - y^n \leq n - 1$$

if $y \geq 0$ for all $n \in \mathbb{N}$.

Problem 24

Topic: Matrices

Problem: For $a, b \in \mathbb{R}$, prove

$$\begin{bmatrix} a & 0 \\ 0 & b \end{bmatrix}^n = \begin{bmatrix} a^n & 0 \\ 0 & b^n \end{bmatrix}$$

for all $n \in \mathbb{N}$.

Definition - Matrix Multiplication: For those who are unfamiliar with matrices, for our purpose here, we only need to know that

$$\begin{bmatrix} a_{11} & a_{12} \\ a_{21} & a_{22} \end{bmatrix} \begin{bmatrix} b_{11} & b_{12} \\ b_{21} & b_{22} \end{bmatrix} = \begin{bmatrix} a_{11}b_{11} + a_{12}b_{21} & a_{11}b_{12} + a_{12}b_{22} \\ a_{21}b_{11} + a_{22}b_{21} & a_{21}b_{12} + a_{22}b_{22} \end{bmatrix}.$$

Solution

Base Cases

We need to prove that it is true for $n = 1$. We can check that

$$\begin{bmatrix} a & 0 \\ 0 & b \end{bmatrix}^1 = \begin{bmatrix} a & 0 \\ 0 & b \end{bmatrix} = \begin{bmatrix} a^1 & 0 \\ 0 & b^1 \end{bmatrix},$$

so it is true for $n = 1$.

Induction Step

Assume

$$\begin{bmatrix} a & 0 \\ 0 & b \end{bmatrix}^k = \begin{bmatrix} a^k & 0 \\ 0 & b^k \end{bmatrix} \tag{24.1}$$

for some $k \geq 1$. To prove the induction step, we need to prove

$$\begin{bmatrix} a & 0 \\ 0 & b \end{bmatrix}^{k+1} = \begin{bmatrix} a^{k+1} & 0 \\ 0 & b^{k+1} \end{bmatrix}.$$

Using assumption (24.1),

$$\begin{aligned}
\begin{bmatrix} a & 0 \\ 0 & b \end{bmatrix}^{k+1} &= \begin{bmatrix} a & 0 \\ 0 & b \end{bmatrix}^{k} \begin{bmatrix} a & 0 \\ 0 & b \end{bmatrix} \\
&= \begin{bmatrix} a^{k} & 0 \\ 0 & b^{k} \end{bmatrix} \begin{bmatrix} a & 0 \\ 0 & b \end{bmatrix} \\
&= \begin{bmatrix} a^{k} \cdot a + 0 \cdot 0 & a^{k} \cdot 0 + 0 \cdot b \\ 0 \cdot a + b^{k} \cdot 0 & 0 \cdot 0 + b^{k} \cdot b \end{bmatrix} \\
&= \begin{bmatrix} a^{k+1} & 0 \\ 0 & b^{k+1} \end{bmatrix}.
\end{aligned}$$

Thus, we proved that it is true for $n = k + 1$ if it is true for $n = k$.

Conclusion

We proved that it is true for $n = 1$. Then, we proved that it is true for $n = k + 1$ if it is true for $n = k$. Therefore, by weak induction,

$$\begin{bmatrix} a & 0 \\ 0 & b \end{bmatrix}^{n} = \begin{bmatrix} a^{n} & 0 \\ 0 & b^{n} \end{bmatrix}$$

for $a, b \in \mathbb{R}$ and for all $n \in \mathbb{N}$.

Problem 25

Topic: Matrices

Problem: For $a \in \mathbb{R}$, prove

$$\begin{bmatrix} 1 & a \\ 0 & 1 \end{bmatrix}^n = \begin{bmatrix} 1 & na \\ 0 & 1 \end{bmatrix}$$

for all $n \in \mathbb{N}$.

Definition - Matrix Multiplication: For those who are unfamiliar with matrices, for our purpose here, we only need to know that

$$\begin{bmatrix} a_{11} & a_{12} \\ a_{21} & a_{22} \end{bmatrix} \begin{bmatrix} b_{11} & b_{12} \\ b_{21} & b_{22} \end{bmatrix} = \begin{bmatrix} a_{11}b_{11} + a_{12}b_{21} & a_{11}b_{12} + a_{12}b_{22} \\ a_{21}b_{11} + a_{22}b_{21} & a_{21}b_{12} + a_{22}b_{22} \end{bmatrix}.$$

Solution

Base Cases

We need to prove that it is true for $n = 1$. We can check that

$$\begin{bmatrix} 1 & a \\ 0 & 1 \end{bmatrix}^1 = \begin{bmatrix} 1 & 1 \cdot a \\ 0 & 1 \end{bmatrix},$$

so it is true for $n = 1$.

Induction Step

Assume

$$\begin{bmatrix} 1 & a \\ 0 & 1 \end{bmatrix}^k = \begin{bmatrix} 1 & ka \\ 0 & 1 \end{bmatrix} \tag{25.1}$$

for some $k \geq 1$. To prove the induction step, we need to prove

$$\begin{bmatrix} 1 & a \\ 0 & 1 \end{bmatrix}^{k+1} = \begin{bmatrix} 1 & (k+1)a \\ 0 & 1 \end{bmatrix}.$$

Using assumption (25.1),

$$\begin{aligned}
\begin{bmatrix} 1 & a \\ 0 & 1 \end{bmatrix}^{k+1} &= \begin{bmatrix} 1 & a \\ 0 & 1 \end{bmatrix}^{k} \begin{bmatrix} 1 & a \\ 0 & 1 \end{bmatrix} \\
&= \begin{bmatrix} 1 & ka \\ 0 & 1 \end{bmatrix} \begin{bmatrix} 1 & a \\ 0 & 1 \end{bmatrix} \\
&= \begin{bmatrix} 1\cdot 1 + ka\cdot 0 & 1\cdot a + ka\cdot 1 \\ 0\cdot 1 + 1\cdot 0 & 0\cdot a + 1\cdot 1 \end{bmatrix} \\
&= \begin{bmatrix} 1 & (k+1)a \\ 0 & 1 \end{bmatrix}.
\end{aligned}$$

Thus, we proved that it is true for $n = k + 1$ if it is true for $n = k$.

Conclusion

We proved that it is true for $n = 1$. Then, we proved that it is true for $n = k + 1$ if it is true for $n = k$. Therefore, by weak induction,

$$\begin{bmatrix} 1 & a \\ 0 & 1 \end{bmatrix}^{n} = \begin{bmatrix} 1 & na \\ 0 & 1 \end{bmatrix}$$

for all $n \in \mathbb{N}$.

Problem 26

Topic: Matrices, Fibonacci Number

Problem: Prove
$$\begin{bmatrix} 1 & 1 \\ 1 & 0 \end{bmatrix}^n = \begin{bmatrix} F_{n+1} & F_n \\ F_n & F_{n-1} \end{bmatrix}$$
for all $n \in \mathbb{N}$ where F_n denotes the n-th Fibonacci number.

Definition - Matrix Multiplication: For those who are unfamiliar with matrices, for our purpose here, we only need to know that
$$\begin{bmatrix} a_{11} & a_{12} \\ a_{21} & a_{22} \end{bmatrix} \begin{bmatrix} b_{11} & b_{12} \\ b_{21} & b_{22} \end{bmatrix} = \begin{bmatrix} a_{11}b_{11} + a_{12}b_{21} & a_{11}b_{12} + a_{12}b_{22} \\ a_{21}b_{11} + a_{22}b_{21} & a_{21}b_{12} + a_{22}b_{22} \end{bmatrix}.$$

Definition - Fibonacci Numbers: The Fibonacci numbers are a sequence of numbers defined as $F_0 = 0$, $F_1 = 1$, and $F_n = F_{n-1} + F_{n-2}$ for $n \geq 2$.

Solution

Base Cases

We need to prove that it is true for $n = 1$. When $n = 1$,

$$\begin{bmatrix} 1 & 1 \\ 1 & 0 \end{bmatrix}^1 = \begin{bmatrix} 1 & 1 \\ 1 & 0 \end{bmatrix} = \begin{bmatrix} F_2 & F_1 \\ F_1 & F_0 \end{bmatrix}$$

because $F_0 = 0$, $F_1 = 1$, and $F_2 = F_1 + F_0 = 1 + 0 = 1$, so it is true for $n = 1$.

Induction Step

Assume

$$\begin{bmatrix} 1 & 1 \\ 1 & 0 \end{bmatrix}^k = \begin{bmatrix} F_{k+1} & F_k \\ F_k & F_{k-1} \end{bmatrix} \qquad (26.1)$$

for some $k \geq 1$. To prove the induction step, we need to prove

$$\begin{bmatrix} 1 & 1 \\ 1 & 0 \end{bmatrix}^{k+1} = \begin{bmatrix} F_{(k+1)+1} & F_{k+1} \\ F_{k+1} & F_{(k+1)-1} \end{bmatrix} = \begin{bmatrix} F_{k+2} & F_{k+1} \\ F_{k+1} & F_k \end{bmatrix}.$$

Using assumption (26.1),

$$\begin{aligned} \begin{bmatrix} 1 & 1 \\ 1 & 0 \end{bmatrix}^{k+1} &= \begin{bmatrix} 1 & 1 \\ 1 & 0 \end{bmatrix}^k \begin{bmatrix} 1 & 1 \\ 1 & 0 \end{bmatrix} \\ &= \begin{bmatrix} F_{k+1} & F_k \\ F_k & F_{k-1} \end{bmatrix} \begin{bmatrix} 1 & 1 \\ 1 & 0 \end{bmatrix} \\ &= \begin{bmatrix} F_{k+1} \cdot 1 + F_k \cdot 1 & F_{k+1} \cdot 1 + F_k \cdot 0 \\ F_k \cdot 1 + F_{k-1} \cdot 1 & F_k \cdot 1 + F_{k-1} \cdot 0 \end{bmatrix} \\ &= \begin{bmatrix} F_k + F_{k+1} & F_{k+1} \\ F_{k-1} + F_k & F_k \end{bmatrix}. \end{aligned}$$

Using the recursive relation

$$F_n = F_{n-1} + F_{n-2}$$

of F_n,

$$F_k + F_{k+1} = F_{k+2}$$

and

$$F_{k-1} + F_k = F_{k+1}.$$

Therefore,

$$\begin{bmatrix} 1 & 1 \\ 1 & 0 \end{bmatrix}^{k+1} = \begin{bmatrix} F_{k+2} & F_{k+1} \\ F_{k+1} & F_k \end{bmatrix}.$$

Thus, we proved that it is true for $n = k + 1$ if it is true for $n = k$.

Conclusion

We proved that it is true for $n = 1$. Then, we proved that it is true for $n = k + 1$ if it is true for $n = k$. Therefore, by weak induction,

$$\begin{bmatrix} 1 & 1 \\ 1 & 0 \end{bmatrix}^n = \begin{bmatrix} F_{n+1} & F_n \\ F_n & F_{n-1} \end{bmatrix}$$

for all $n \in \mathbb{N}$.

Problem 27

Topic: Floor Function, Inequality

Problem: Prove
$$0 \le \left\lfloor \frac{n+1}{2} \right\rfloor - \left\lfloor \frac{n}{2} \right\rfloor \le 1$$
for all $n \in \mathbb{N}$ where $\lfloor \cdot \rfloor$ denotes the floor function.

Definition - Floor Function: For $x \in \mathbb{R}$, the floor function $\lfloor x \rfloor$ is the largest integer not exceeding x. For example, $\lfloor 2 \rfloor = 2$, $\lfloor 3.6 \rfloor = 3$, $\lfloor -1.2 \rfloor = -2$, etc. Formally, for any $\ell \in \mathbb{Z}$,

$$\lfloor x \rfloor = \ell \Leftrightarrow \ell \le x < \ell + 1.$$

With this definition of floor function, it should be easy to show that $\lfloor x \rfloor$ has the property
$$\lfloor x + a \rfloor = \lfloor x \rfloor + a$$
for any $a \in \mathbb{Z}$.

Solution

Base Cases

We need to prove that it is true for $n = 1$. When $n = 1$,

$$\left\lfloor \frac{1+1}{2} \right\rfloor - \left\lfloor \frac{1}{2} \right\rfloor = 1 - 0 = 1,$$

so it is true that

$$0 \le \left\lfloor \frac{1+1}{2} \right\rfloor - \left\lfloor \frac{1}{2} \right\rfloor \le 1.$$

Induction Step

Assume

$$0 \leq \left\lfloor \frac{k+1}{2} \right\rfloor - \left\lfloor \frac{k}{2} \right\rfloor \leq 1 \tag{27.1}$$

for some $k \geq 1$. To prove the induction step, we need to prove

$$0 \leq \left\lfloor \frac{(k+1)+1}{2} \right\rfloor - \left\lfloor \frac{k+1}{2} \right\rfloor \leq 1.$$

Note that

$$\left\lfloor \frac{(k+1)+1}{2} \right\rfloor - \left\lfloor \frac{k+1}{2} \right\rfloor = \left\lfloor \frac{k+2}{2} \right\rfloor - \left\lfloor \frac{k+1}{2} \right\rfloor$$

$$= \left\lfloor \frac{k}{2} + 1 \right\rfloor - \left\lfloor \frac{k+1}{2} \right\rfloor$$

$$= \left\lfloor \frac{k}{2} \right\rfloor + 1 - \left\lfloor \frac{k+1}{2} \right\rfloor$$

$$= 1 - \left(\left\lfloor \frac{k+1}{2} \right\rfloor - \left\lfloor \frac{k}{2} \right\rfloor \right).$$

By assumption (27.1),

$$0 \leq \left\lfloor \frac{k+1}{2} \right\rfloor - \left\lfloor \frac{k}{2} \right\rfloor \leq 1,$$

so

$$0 \leq 1 - \left(\left\lfloor \frac{k+1}{2} \right\rfloor - \left\lfloor \frac{k}{2} \right\rfloor \right) \leq 1.$$

Therefore,

$$0 \leq \left\lfloor \frac{(k+1)+1}{2} \right\rfloor - \left\lfloor \frac{k+1}{2} \right\rfloor \leq 1.$$

Thus, we proved that it is true for $n = k+1$ if it is true for $n = k$.

Conclusion

We proved that it is true for $n = 1$. Then, we proved that it is true for $n = k+1$ if it is true for $n = k$. Therefore, by weak induction,

$$0 \leq \left\lfloor \frac{n+1}{2} \right\rfloor - \left\lfloor \frac{n}{2} \right\rfloor \leq 1$$

for all $n \in \mathbb{N}$.

Problem 28

Topic: Floor Function, Inequality

Problem: Prove

$$\left\lfloor \frac{n+1}{2} \right\rfloor \geq \frac{n}{2}$$

for all $n \in \mathbb{N}$ where $\lfloor \cdot \rfloor$ denotes the floor function.

Definition - Floor Function: For $x \in \mathbb{R}$, the floor function $\lfloor x \rfloor$ is the largest integer not exceeding x. For example, $\lfloor 2 \rfloor = 2$, $\lfloor 3.6 \rfloor = 3$, $\lfloor -1.2 \rfloor = -2$, etc. Formally, for any $\ell \in \mathbb{Z}$,

$$\lfloor x \rfloor = \ell \Leftrightarrow \ell \leq x < \ell + 1.$$

With this definition of floor function, it should be easy to show that $\lfloor x \rfloor$ has the property

$$\lfloor x + a \rfloor = \lfloor x \rfloor + a$$

for any $a \in \mathbb{Z}$.

Solution

Base Cases

We need to prove that it is true for $n = 1$ and $n = 2$. We can check that

$$\left\lfloor \frac{1+1}{2} \right\rfloor = 1 \geq \frac{1}{2}$$

and

$$\left\lfloor \frac{2+1}{2} \right\rfloor = 1 \geq \frac{2}{2},$$

so it is true for $n = 1$ and $n = 2$.

Induction Step

Assume

$$\left\lfloor \frac{m+1}{2} \right\rfloor \geq \frac{m}{2} \tag{28.1}$$

for all $1 \leq m \leq k$ for some $k \geq 2$. To prove the induction step, we need to prove

$$\left\lfloor \frac{(k+1)+1}{2} \right\rfloor \geq \frac{k+1}{2}.$$

Note that

$$\left\lfloor \frac{(k+1)+1}{2} \right\rfloor = \left\lfloor \frac{k+2}{2} \right\rfloor = \left\lfloor \frac{k}{2} + 1 \right\rfloor = \left\lfloor \frac{k}{2} \right\rfloor + 1.$$

By assumption (28.1) when $m = k - 1$,

$$\left\lfloor \frac{(k-1)+1}{2} \right\rfloor = \left\lfloor \frac{k}{2} \right\rfloor \geq \frac{k-1}{2}.$$

Therefore,

$$\left\lfloor \frac{(k+1)+1}{2} \right\rfloor \geq \frac{k-1}{2} + 1 = \frac{k+1}{2}.$$

Thus, we proved that it is true for $n = k+1$ if it is true for $n = m$ for all $1 \leq m \leq k$.

Conclusion

We proved that it is true for $n = 1$ and $n = 2$. Then, we proved that it is true for $n = k+1$ if it is true for $n = m$ for all $1 \leq m \leq k$. Therefore, by strong induction,

$$\left\lfloor \frac{n+1}{2} \right\rfloor \geq \frac{n}{2}$$

for all $n \in \mathbb{N}$.

Problem 29

Topic: Floor Function, Inequality

Problem: Prove

$$\lfloor \sqrt{n} \rfloor \leq \frac{n}{2}$$

for $n \geq 2$ where $\lfloor \cdot \rfloor$ denotes the floor function.

Definition - Floor Function: For $x \in \mathbb{R}$, the floor function $\lfloor x \rfloor$ is the largest integer not exceeding x. For example, $\lfloor 2 \rfloor = 2$, $\lfloor 3.6 \rfloor = 3$, $\lfloor -1.2 \rfloor = -2$, etc. Formally, for any $\ell \in \mathbb{Z}$,

$$\lfloor x \rfloor = \ell \Leftrightarrow \ell \leq x < \ell + 1.$$

With this definition of floor function, it should be easy to show that $\lfloor x \rfloor$ has the property

$$\lfloor x + a \rfloor = \lfloor x \rfloor + a$$

for any $a \in \mathbb{Z}$.

Solution

Base Cases

We need to prove that it is true for $n = 2$ and $n = 3$. When $n = 2$,

$$\sqrt{1} < \sqrt{2} < \sqrt{4} \Leftrightarrow 1 < \sqrt{2} < 2,$$

so

$$\lfloor \sqrt{2} \rfloor = 1 \leq \frac{2}{2}.$$

When $n = 3$,

$$\sqrt{1} < \sqrt{3} < \sqrt{4} \Leftrightarrow 1 < \sqrt{3} < 2,$$

so
$$\left\lfloor \sqrt{3} \right\rfloor = 1 \leq \frac{3}{2}.$$

Induction Step

Assume
$$\left\lfloor \sqrt{m} \right\rfloor \leq \frac{m}{2} \tag{29.1}$$

for all $2 \leq m \leq k$ for some $k \geq 3$. To prove the induction step, we need to prove
$$\left\lfloor \sqrt{k+1} \right\rfloor \leq \frac{k+1}{2}.$$

Note that for $k \geq 3$,
$$2\sqrt{k-1} \geq 2\sqrt{2} \geq 1,$$

so
$$k + 2\sqrt{k-1} \geq k+1.$$

Since
$$k + 2\sqrt{k-1} = (k-1) + 1 + 2\sqrt{k-1} = \left(\sqrt{k-1}+1\right)^2,$$

we have the inequality
$$\left(\sqrt{k-1}+1\right)^2 \geq k+1.$$

Taking square root of both sides,
$$\sqrt{k-1}+1 \geq \sqrt{k+1}.$$

So,
$$\left\lfloor \sqrt{k+1} \right\rfloor \leq \left\lfloor \sqrt{k-1}+1 \right\rfloor = \left\lfloor \sqrt{k-1} \right\rfloor + 1.$$

By assumption (29.1) when $m = k-1$,
$$\left\lfloor \sqrt{k-1} \right\rfloor \leq \frac{k-1}{2}.$$

Therefore,
$$\left\lfloor \sqrt{k+1} \right\rfloor \leq \frac{k-1}{2} + 1 = \frac{k+1}{2}.$$

Thus, we proved that it is true for $n = k+1$ if it is true for $n = m$ for all $2 \leq m \leq k$.

Conclusion

We proved that it is true for $n = 2$ and $n = 3$. Then, we proved that it is true for $n = k+1$ if it is true for $n = m$ for all $2 \leq m \leq k$. Therefore, by strong induction,
$$\left\lfloor \sqrt{n} \right\rfloor \leq \frac{n}{2}$$

for $n \geq 2$.

Problem 30

Topic: Floor Function

Problem: Prove a floor function identity of Ramanujan,

$$\left\lfloor \frac{n}{3} \right\rfloor + \left\lfloor \frac{n+2}{6} \right\rfloor + \left\lfloor \frac{n+4}{6} \right\rfloor = \left\lfloor \frac{n}{2} \right\rfloor + \left\lfloor \frac{n+3}{6} \right\rfloor,$$

for all $n \in \mathbb{N}$ where $\lfloor \cdot \rfloor$ denotes the floor function.

Definition - Floor Function: For $x \in \mathbb{R}$, the floor function $\lfloor x \rfloor$ is the largest integer not exceeding x. For example, $\lfloor 2 \rfloor = 2$, $\lfloor 3.6 \rfloor = 3$, $\lfloor -1.2 \rfloor = -2$, etc. Formally, for any $\ell \in \mathbb{Z}$,

$$\lfloor x \rfloor = \ell \Leftrightarrow \ell \leq x < \ell + 1.$$

With this definition of floor function, it should be easy to show that $\lfloor x \rfloor$ has the property

$$\lfloor x + a \rfloor = \lfloor x \rfloor + a$$

for any $a \in \mathbb{Z}$.

Solution

Base Cases

We need to prove that it is true for $n = 1$, $n = 2$, $n = 3$, $n = 4$, $n = 5$, and $n = 6$. We can check that

$$\left\lfloor \frac{1}{3} \right\rfloor + \left\lfloor \frac{1+2}{6} \right\rfloor + \left\lfloor \frac{1+4}{6} \right\rfloor = \left\lfloor \frac{1}{2} \right\rfloor + \left\lfloor \frac{1+3}{6} \right\rfloor = 0,$$

$$\left\lfloor \frac{2}{3} \right\rfloor + \left\lfloor \frac{2+2}{6} \right\rfloor + \left\lfloor \frac{2+4}{6} \right\rfloor = \left\lfloor \frac{2}{2} \right\rfloor + \left\lfloor \frac{2+3}{6} \right\rfloor = 1,$$

$$\left\lfloor \frac{3}{3} \right\rfloor + \left\lfloor \frac{3+2}{6} \right\rfloor + \left\lfloor \frac{3+4}{6} \right\rfloor = \left\lfloor \frac{3}{2} \right\rfloor + \left\lfloor \frac{3+3}{6} \right\rfloor = 2,$$

$$\left\lfloor \frac{4}{3} \right\rfloor + \left\lfloor \frac{4+2}{6} \right\rfloor + \left\lfloor \frac{4+4}{6} \right\rfloor = \left\lfloor \frac{4}{2} \right\rfloor + \left\lfloor \frac{4+3}{6} \right\rfloor = 3,$$

$$\left\lfloor \frac{5}{3} \right\rfloor + \left\lfloor \frac{5+2}{6} \right\rfloor + \left\lfloor \frac{5+4}{6} \right\rfloor = \left\lfloor \frac{5}{2} \right\rfloor + \left\lfloor \frac{5+3}{6} \right\rfloor = 3,$$

and

$$\left\lfloor \frac{6}{3} \right\rfloor + \left\lfloor \frac{6+2}{6} \right\rfloor + \left\lfloor \frac{6+4}{6} \right\rfloor = \left\lfloor \frac{6}{2} \right\rfloor + \left\lfloor \frac{6+3}{6} \right\rfloor = 4,$$

so it is true for $n = 1$, $n = 2$, $n = 3$, $n = 4$, $n = 5$, and $n = 6$.

Induction Step

Assume

$$\left\lfloor \frac{m}{3} \right\rfloor + \left\lfloor \frac{m+2}{6} \right\rfloor + \left\lfloor \frac{m+4}{6} \right\rfloor = \left\lfloor \frac{m}{2} \right\rfloor + \left\lfloor \frac{m+3}{6} \right\rfloor \qquad (30.1)$$

for all $1 \leq m \leq k$ for some $k \geq 6$. To prove the induction step, we need to prove

$$\left\lfloor \frac{k+1}{3} \right\rfloor + \left\lfloor \frac{(k+1)+2}{6} \right\rfloor + \left\lfloor \frac{(k+1)+4}{6} \right\rfloor = \left\lfloor \frac{k+1}{2} \right\rfloor + \left\lfloor \frac{(k+1)+3}{6} \right\rfloor.$$

Note that

$$\left\lfloor \frac{k+1}{3} \right\rfloor + \left\lfloor \frac{(k+1)+2}{6} \right\rfloor + \left\lfloor \frac{(k+1)+4}{6} \right\rfloor$$

$$= \left\lfloor \frac{k-5}{3} + 2 \right\rfloor + \left\lfloor \frac{(k-5)+2}{6} + 1 \right\rfloor + \left\lfloor \frac{(k-5)+4}{6} + 1 \right\rfloor$$

$$= \left\lfloor \frac{k-5}{3} \right\rfloor + 2 + \left\lfloor \frac{(k-5)+2}{6} \right\rfloor + 1 + \left\lfloor \frac{(k-5)+4}{6} \right\rfloor + 1$$

$$= \left\lfloor \frac{k-5}{3} \right\rfloor + \left\lfloor \frac{(k-5)+2}{6} \right\rfloor + \left\lfloor \frac{(k-5)+4}{6} \right\rfloor + 4.$$

By assumption (30.1) when $m = k - 5$,

$$\left\lfloor \frac{k-5}{3} \right\rfloor + \left\lfloor \frac{(k-5)+2}{6} \right\rfloor + \left\lfloor \frac{(k-5)+4}{6} \right\rfloor = \left\lfloor \frac{k-5}{2} \right\rfloor + \left\lfloor \frac{(k-5)+3}{6} \right\rfloor.$$

Therefore,

$$\left\lfloor \frac{k+1}{3} \right\rfloor + \left\lfloor \frac{(k+1)+2}{6} \right\rfloor + \left\lfloor \frac{(k+1)+4}{6} \right\rfloor$$

$$= \left\lfloor \frac{k-5}{2} \right\rfloor + \left\lfloor \frac{(k-5)+3}{6} \right\rfloor + 4$$

$$= \left\lfloor \frac{k-5}{2} \right\rfloor + 3 + \left\lfloor \frac{(k-5)+3}{6} \right\rfloor + 1$$

$$= \left\lfloor \frac{k-5}{2} + 3 \right\rfloor + \left\lfloor \frac{(k-5)+3}{6} + 1 \right\rfloor$$

$$= \left\lfloor \frac{k+1}{2} \right\rfloor + \left\lfloor \frac{(k+1)+3}{6} \right\rfloor.$$

Thus, we proved that it is true for $n = k+1$ if it is true for $n = m$ for all $1 \le m \le k$.

Conclusion

We proved that it is true for $n = 1$, $n = 2$, $n = 3$, $n = 4$, $n = 5$, and $n = 6$. Then, we proved that it is true for $n = k + 1$ if it is true for $n = m$ for all $1 \le m \le k$. Therefore, by strong induction,

$$\left\lfloor \frac{n}{3} \right\rfloor + \left\lfloor \frac{n+2}{6} \right\rfloor + \left\lfloor \frac{n+4}{6} \right\rfloor = \left\lfloor \frac{n}{2} \right\rfloor + \left\lfloor \frac{n+3}{6} \right\rfloor$$

for all $n \in \mathbb{N}$.

Problem 31

Topic: Sequence

Problem: Let a_1, a_2 be arbitrary real numbers, and

$$a_n = 2a_{n-1} - a_{n-2}$$

for $n \geq 3$. Prove that

$$a_n = (n-1)a_2 - (n-2)a_1$$

for all $n \in \mathbb{N}$.

Solution

Base Cases

We need to prove that it is true for $n = 1$ and $n = 2$. We can check that

$$(1-1)a_2 - (1-2)a_1 = a_1$$

and

$$(2-1)a_2 - (2-2)a_1 = a_2,$$

so it is true for $n = 1$ and $n = 2$.

Induction Step

Assume

$$a_m = (m-1)a_2 - (m-2)a_1 \tag{31.1}$$

for all $1 \leq m \leq k$ for some $k \geq 2$. To prove the induction step, we need to prove

$$a_{k+1} = ((k+1)-1)a_2 - ((k+1)-2)a_1 = ka_2 - (k-1)a_1.$$

Since $k + 1 \geq 3$, we can use the recursive formula and get

$$a_{k+1} = 2a_k - a_{k-1}.$$

By assumption (31.1) when $m = k$,

$$a_k = (k - 1)a_2 - (k - 2)a_1.$$

By assumption (31.1) when $m = k - 1$,

$$a_{k-1} = (k - 2)a_2 - (k - 3)a_1.$$

So,

$$\begin{aligned}
a_{k+1} &= 2[(k - 1)a_2 - (k - 2)a_1] - [(k - 2)a_2 - (k - 3)a_1] \\
&= ka_2 - (k - 1)a_1.
\end{aligned}$$

Thus, we proved that it is true for $n = k+1$ if it is true for $n = m$ for all $1 \leq m \leq k$.

Conclusion

We proved that it is true for $n = 1$ and $n = 2$. Then, we proved that it is true for $n = k + 1$ if it is true for $n = m$ for all $1 \leq m \leq k$. Therefore, by strong induction,

$$a_n = (n - 1)a_2 - (n - 2)a_1$$

for all $n \in \mathbb{N}$.

Problem 32

Topic: Sequence

Problem: Let $a_1 = 5$, $a_2 = 11$, $a_3 = 29$, and

$$a_n = 3a_{n-1} + a_{n-2} - 3a_{n-3}$$

for $n \geq 4$. Prove that

$$a_n = 3^n + 2$$

for all $n \in \mathbb{N}$.

Solution

Base Cases

We need to prove that it is true for $n = 1$, $n = 2$, and $n = 3$. We can check that

$$a_1 = 5 = 3^1 + 2,$$

$$a_2 = 11 = 3^2 + 2,$$

and

$$a_3 = 29 = 3^3 + 2,$$

so it is true for $n = 1$, $n = 2$, and $n = 3$.

Induction Step

Assume

$$a_m = 3^m + 2 \tag{32.1}$$

for all $1 \leq m \leq k$ for some $k \geq 3$. To prove the induction step, we need to prove

$$a_{k+1} = 3^{k+1} + 2.$$

Since $k + 1 \geq 4$, we can use the recursive formula and get

$$a_{k+1} = 3a_k + a_{k-1} - 3a_{k-2}.$$

By assumption (32.1) when $m = k$,

$$a_k = 3^k + 2.$$

By assumption (32.1) when $m = k - 1$,

$$a_{k-1} = 3^{k-1} + 2.$$

By assumption (32.1) when $m = k - 2$,

$$a_{k-2} = 3^{k-2} + 2.$$

So,

$$\begin{aligned}
a_{k+1} &= 3\left(3^k + 2\right) + \left(3^{k-1} + 2\right) - 3\left(3^{k-2} + 2\right) \\
&= 3^{k+1} + 2.
\end{aligned}$$

Thus, we proved that it is true for $n = k+1$ if it is true for $n = m$ for all $1 \leq m \leq k$.

Conclusion

We proved that it is true for $n = 1$, $n = 2$, and $n = 3$. Then, we proved that it is true for $n = k + 1$ if it is true for $n = m$ for all $1 \leq m \leq k$. Therefore, by strong induction,

$$a_n = 3^n + 2$$

for all $n \in \mathbb{N}$.

Problem 33

Topic: Sequence

Problem: Let a_1, a_2, a_3 be arbitrary real numbers, and

$$a_n = 3a_{n-1} - 3a_{n-2} + a_{n-3}$$

for $n \geq 4$. Prove that

$$a_n = \frac{1}{2}(n-1)(n-2)a_3 - (n-1)(n-3)a_2 + \frac{1}{2}(n-2)(n-3)a_1$$

for all $n \in \mathbb{N}$.

Solution

Base Cases

We need to prove that it is true for $n = 1$, $n = 2$, and $n = 3$. We can check that

$$\frac{1}{2}(1-1)(1-2)a_3 - (1-1)(1-3)a_2 + \frac{1}{2}(1-2)(1-3)a_1 = a_1,$$

$$\frac{1}{2}(2-1)(2-2)a_3 - (2-1)(2-3)a_2 + \frac{1}{2}(2-2)(2-3)a_1 = a_2,$$

and

$$\frac{1}{2}(3-1)(3-2)a_3 - (3-1)(3-3)a_2 + \frac{1}{2}(3-2)(3-3)a_1 = a_3,$$

so it is true for $n = 1$, $n = 2$, and $n = 3$.

Induction Step

Assume

$$a_m = \frac{1}{2}(m-1)(m-2)a_3 - (m-1)(m-3)a_2 + \frac{1}{2}(m-2)(m-3)a_1 \qquad (33.1)$$

for all $1 \leq m \leq k$ for some $k \geq 3$. To prove the induction step, we need to prove

$$a_{k+1} = \frac{1}{2}k(k-1)a_3 - k(k-2)a_2 + \frac{1}{2}(k-1)(k-2)a_1.$$

Since $k + 1 \geq 4$, we can use the recursive formula and get

$$a_{k+1} = 3a_k - 3a_{k-1} + a_{k-2}.$$

By assumption (33.1) when $m = k$,

$$a_k = \frac{1}{2}(k-1)(k-2)a_3 - (k-1)(k-3)a_2 + \frac{1}{2}(k-2)(k-3)a_1.$$

By assumption (33.1) when $m = k - 1$,

$$a_{k-1} = \frac{1}{2}(k-2)(k-3)a_3 - (k-2)(k-4)a_2 + \frac{1}{2}(k-3)(k-4)a_1.$$

By assumption (33.1) when $m = k - 2$,

$$a_{k-2} = \frac{1}{2}(k-3)(k-4)a_3 - (k-3)(k-5)a_2 + \frac{1}{2}(k-4)(k-5)a_1.$$

So,

$$
\begin{aligned}
a_{k+1} =& \frac{3}{2}(k-1)(k-2)a_3 - 3(k-1)(k-3)a_2 + \frac{3}{2}(k-2)(k-3)a_1 \\
& - \frac{3}{2}(k-2)(k-3)a_3 + 3(k-2)(k-4)a_2 - \frac{3}{2}(k-3)(k-4)a_1 \\
& + \frac{1}{2}(k-3)(k-4)a_3 - (k-3)(k-5)a_2 + \frac{1}{2}(k-4)(k-5)a_1 \\
=& \frac{1}{2}k(k-1)a_3 - k(k-2)a_2 + \frac{1}{2}(k-1)(k-2)a_1.
\end{aligned}
$$

Thus, we proved that it is true for $n = k+1$ if it is true for $n = m$ for all $1 \leq m \leq k$.

Conclusion

We proved that it is true for $n = 1$, $n = 2$, and $n = 3$. Then, we proved that it is true for $n = k + 1$ if it is true for $n = m$ for all $1 \leq m \leq k$. Therefore, by strong induction,

$$a_n = \frac{1}{2}(n-1)(n-2)a_3 - (n-1)(n-3)a_2 + \frac{1}{2}(n-2)(n-3)a_1$$

for all $n \in \mathbb{N}$.

Problem 34

Topic: Sequence, Inequality

Problem: Let $a_0 = 1$ and $a_n = na_{n-1} + n$ for $n \geq 1$. Prove that

$$a_n \geq 3 \cdot n!$$

for $n \geq 2$.

Definition - Factorial:

$$n! = n \cdot (n-1) \cdot (n-2) \cdots 2 \cdot 1.$$

$$0! = 1.$$

Solution

Base Cases

We need to prove that it is true for $n = 2$. Using the recursive formula,

$$a_1 = 1a_0 + 1 = 1 \cdot 1 + 1 = 2,$$

and

$$a_2 = 2a_1 + 2 = 2 \cdot 2 + 2 = 6 \geq 3 \cdot 2!,$$

so it is true for $n = 2$.

Induction Step

Assume

$$a_k \geq 3 \cdot k! \tag{34.1}$$

for some $k \geq 2$. To prove the induction step, we need to prove

$$a_{k+1} \geq 3 \cdot (k+1)!.$$

Using the recursive formula and by assumption (34.1),

$$\begin{aligned}
a_{k+1} &= (k+1)a_k + (k+1) \\
&\geq (k+1) \cdot 3 \cdot k! + k + 1 \\
&\geq (k+1) \cdot 3 \cdot k! \\
&= 3 \cdot (k+1)!.
\end{aligned}$$

Thus, we proved that it is true for $n = k+1$ if it is true for $n = k$.

Conclusion

We proved that it is true for $n = 2$. Then, we proved that it is true for $n = k+1$ if it is true for $n = k$. Therefore, by weak induction,

$$a_n \geq 3 \cdot n!$$

for $n \geq 2$.

Problem 35

Topic: Sequence, Inequality

Problem: Let $a_0 = 1$ and $a_n = na_{n-1} + r$ for $n \geq 1$ where r is some non-negative real number. Prove that

$$a_n \geq (1 + r) \cdot n!$$

for all $n \in \mathbb{N}$.

Definition - Factorial:

$$n! = n \cdot (n - 1) \cdot (n - 2) \cdots 2 \cdot 1.$$

$$0! = 1.$$

Solution

Base Cases

We need to prove that it is true for $n = 1$. We can check that

$$a_1 = 1a_0 + r = 1 \cdot 1 + r = 1 + r \geq (1 + r) \cdot 1!,$$

so it is true for $n = 1$.

Induction Step

Assume

$$a_k \geq (1 + r) \cdot k! \tag{35.1}$$

for some $k \geq 1$. To prove the induction step, we need to prove

$$a_{k+1} \geq (1 + r) \cdot (k + 1)!.$$

Using the recursive formula and by assumption (35.1), since r is non-negative,

$$\begin{aligned}
a_{k+1} &= (k + 1)a_k + r \\
&\geq (k + 1) \cdot (1 + r) \cdot k! + r \\
&\geq (k + 1) \cdot (1 + r) \cdot k! \\
&= (1 + r) \cdot (k + 1)!.
\end{aligned}$$

Thus, we proved that it is true for $n = k + 1$ if it is true for $n = k$.

Conclusion

We proved that it is true for $n = 1$. Then, we proved that it is true for $n = k + 1$ if it is true for $n = k$. Therefore, by weak induction,

$$a_n \geq (1 + r) \cdot n!$$

for all $n \in \mathbb{N}$.

Problem 36

Topic: Sequence, Inequality

Problem: Let a_0 be an arbitrary real number such that $0 < a_0 < 1$. Let

$$a_n = \sqrt{\frac{1 + a_{n-1}}{2}}$$

for $n \geq 1$. Prove that

$$1 - \frac{1}{2^n} < a_n < 1$$

for all $n \in \mathbb{N}$.

Solution

Base Cases

We need to prove that it is true for $n = 1$. Using the recursive formula,

$$a_1 = \sqrt{\frac{1 + a_0}{2}}.$$

Since $0 < a_0 < 1$,

$$1 < 1 + a_0 < 2$$
$$\frac{1}{2} < \frac{1 + a_0}{2} < 1$$
$$\frac{1}{\sqrt{2}} < \sqrt{\frac{1 + a_0}{2}} < 1.$$

Since $\sqrt{2} < \sqrt{4} = 2$,

$$\frac{1}{\sqrt{2}} > \frac{1}{2}.$$

So,
$$1 - \frac{1}{2^1} = \frac{1}{2} < a_1 < 1,$$
and it is true for $n = 1$.

Induction Step

Assume
$$1 - \frac{1}{2^k} < a_k < 1 \tag{36.1}$$
for some $k \geq 1$. To prove the induction step, we need to prove
$$1 - \frac{1}{2^{k+1}} < a_{k+1} < 1.$$

Using the recursive formula,
$$a_{k+1} = \sqrt{\frac{1 + a_k}{2}}.$$

By assumption (36.1),
$$2 - \frac{1}{2^k} < 1 + a_k < 2$$
$$1 - \frac{1}{2^{k+1}} < \frac{1 + a_k}{2} < 1$$
$$\sqrt{1 - \frac{1}{2^{k+1}}} < \sqrt{\frac{1 + a_k}{2}} < 1,$$
so
$$\sqrt{1 - \frac{1}{2^{k+1}}} < a_{k+1} < 1.$$

Since $1 - \frac{1}{2^{k+1}} < 1$,
$$\sqrt{1 - \frac{1}{2^{k+1}}} < 1,$$
and multiplying $\sqrt{1 - \frac{1}{2^{k+1}}}$ to both sides gives
$$1 - \frac{1}{2^{k+1}} < \sqrt{1 - \frac{1}{2^{k+1}}}.$$

So,
$$1 - \frac{1}{2^{k+1}} < a_{k+1} < 1.$$
Thus, we proved that it is true for $n = k + 1$ if it is true for $n = k$.

Conclusion

We proved that it is true for $n = 1$. Then, we proved that it is true for $n = k + 1$ if it is true for $n = k$. Therefore, by weak induction,
$$1 - \frac{1}{2^n} < a_n < 1$$
for all $n \in \mathbb{N}$.

Problem 37

Topic: Sequence

Problem: Let a_0 be an arbitrary real number, and

$$a_n = 2a_{n-1}^2$$

for $n \geq 1$. Prove that

$$a_n = 2^{2^n - 1} a_0^{2^n}$$

for all $n \in \mathbb{N}$.

Solution

Base Cases

We need to prove that it is true for $n = 1$. Using the recursive formula,

$$a_1 = 2a_0^2.$$

We can check that

$$2^{2^1 - 1} a_0^{2^1} = 2a_0^2 = a_1,$$

so it is true for $n = 1$.

Induction Step

Assume

$$a_k = 2^{2^k - 1} a_0^{2^k} \tag{37.1}$$

for some $k \geq 1$. To prove the induction step, we need to prove

$$a_{k+1} = 2^{2^{k+1} - 1} a_0^{2^{k+1}}.$$

Using the recursive formula and the assumption (37.1),

$$a_{k+1} = 2a_k^2$$
$$= 2\left(2^{2^k-1}a_0^{2^k}\right)^2$$
$$= 2 \cdot 2^{2\cdot 2^k - 2}a_0^{2\cdot 2^k}$$
$$= 2^{2^{k+1}-1}a_0^{2^{k+1}}.$$

Thus, we proved that it is true for $n = k+1$ if it is true for $n = k$.

Conclusion

We proved that it is true for $n = 1$. Then, we proved that it is true for $n = k+1$ if it is true for $n = k$. Therefore, by weak induction,

$$a_n = 2^{2^n-1}a_0^{2^n}$$

for all $n \in \mathbb{N}$.

Problem 38

Topic: Sequence

Problem: Let $a_0 = -2$ and

$$a_n = \frac{(n+2)a_{n-1}}{n+1+2a_{n-1}}$$

for $n \geq 1$. Prove that

$$a_n = \frac{n+2}{2n-1}$$

for all $n \in \mathbb{N}$.

Solution

Base Cases

We need to prove that it is true for $n = 1$. By the recursive formula of a_n,

$$a_1 = \frac{(1+2)a_0}{1+1+2a_0} = \frac{3 \cdot (-2)}{2 + 2 \cdot (-2)} = 3 = \frac{1+2}{2 \cdot 1 - 1},$$

so it is true for $n = 1$.

Induction Step

Assume

$$a_k = \frac{k+2}{2k-1} \tag{38.1}$$

for some $k \geq 1$. To prove the induction step, we need to prove

$$a_{k+1} = \frac{(k+1)+2}{2(k+1)-1} = \frac{k+3}{2k+1}.$$

Using the recursive formula of a_n and the assumption (38.1),

$$\begin{aligned}
a_{k+1} &= \frac{(k+3)a_k}{k+2+2a_k} \\
&= \frac{(k+3)\cdot\frac{k+2}{2k-1}}{k+2+2\cdot\frac{k+2}{2k-1}} \\
&= \frac{(k+3)(k+2)}{(k+2)(2k-1)+2(k+2)} \\
&= \frac{(k+3)(k+2)}{(k+2)(2k+1)} \\
&= \frac{k+3}{2k+1}.
\end{aligned}$$

Thus, we proved that it is true for $n = k+1$ if it is true for $n = k$.

Conclusion

We proved that it is true for $n = 1$. Then, we proved that it is true for $n = k+1$ if it is true for $n = k$. Therefore, by weak induction,

$$a_n = \frac{n+2}{2n-1}$$

for all $n \in \mathbb{N}$.

Problem 39

Topic: Sequence, Floor Function, Inequality

Problem: Let a_0 be an arbitrary real number, and

$$a_n = \frac{\lfloor a_{n-1} \rfloor}{2} + \frac{a_{n-1}}{2}$$

for $n \geq 1$ where $\lfloor \cdot \rfloor$ denotes the floor function. Prove that

$$a_0 - \frac{n}{2} < a_n \leq a_0$$

for all $n \in \mathbb{N}$.

Definition - Floor Function: For $x \in \mathbb{R}$, the floor function $\lfloor x \rfloor$ is the largest integer not exceeding x. For example, $\lfloor 2 \rfloor = 2$, $\lfloor 3.6 \rfloor = 3$, $\lfloor -1.2 \rfloor = -2$, etc. Formally, for any $\ell \in \mathbb{Z}$,

$$\lfloor x \rfloor = \ell \Leftrightarrow \ell \leq x < \ell + 1.$$

With this definition of floor function, it should be easy to show that $\lfloor x \rfloor$ has the property
$$\lfloor x + a \rfloor = \lfloor x \rfloor + a$$

for any $a \in \mathbb{Z}$.

Solution

Base Cases

We need to prove that it is true for $n = 1$. By the recursive formula,

$$a_1 = \frac{\lfloor a_0 \rfloor}{2} + \frac{a_0}{2}.$$

From the definition of floor function, we know that

$$\lfloor a_0 \rfloor \le a_0 < \lfloor a_0 \rfloor + 1 \Leftrightarrow a_0 - 1 < \lfloor a_0 \rfloor \le a_0,$$

so

$$\frac{a_0 - 1}{2} < \frac{\lfloor a_0 \rfloor}{2} \le \frac{a_0}{2}$$

$$a_0 - \frac{1}{2} < \frac{\lfloor a_0 \rfloor}{2} + \frac{a_0}{2} \le a_0.$$

Hence,

$$a_0 - \frac{1}{2} < a_1 \le a_0,$$

and it is true for $n = 1$.

Induction Step

Assume

$$a_0 - \frac{k}{2} < a_k \le a_0 \tag{39.1}$$

for some $k \ge 1$. To prove the induction step, we need to prove

$$a_0 - \frac{k+1}{2} < a_{k+1} \le a_0.$$

By the recursive formula,

$$a_{k+1} = \frac{\lfloor a_k \rfloor}{2} + \frac{a_k}{2}.$$

From the definition of floor function, we know that

$$\lfloor a_k \rfloor \le a_k < \lfloor a_k \rfloor + 1 \Leftrightarrow a_k - 1 < \lfloor a_k \rfloor \le a_k,$$

so

$$\frac{a_k - 1}{2} < \frac{\lfloor a_k \rfloor}{2} \le \frac{a_k}{2}$$

$$a_k - \frac{1}{2} < \frac{\lfloor a_k \rfloor}{2} + \frac{a_k}{2} \le a_k,$$

and we have

$$a_k - \frac{1}{2} < a_{k+1} \le a_k.$$

Using assumption (39.1),

$$a_{k+1} \le a_k \le a_0,$$

and

$$a_{k+1} > a_k - \frac{1}{2} > a_0 - \frac{k}{2} - \frac{1}{2} = a_0 - \frac{k+1}{2}.$$

Hence,

$$a_0 - \frac{k+1}{2} < a_{k+1} \le a_0.$$

Thus, we proved that it is true for $n = k + 1$ if it is true for $n = k$.

Conclusion

We proved that it is true for $n = 1$. Then, we proved that it is true for $n = k + 1$ if it is true for $n = k$. Therefore, by weak induction,

$$a_0 - \frac{n}{2} < a_n \leq a_0$$

for all $n \in \mathbb{N}$.

Problem 40

Topic: Sequence, Floor Function, Inequality

Problem: Let a_1, b_1 be arbitrary real numbers such that $a_1 > b_1$. Let

$$a_n = \frac{\lfloor a_{n-1} \rfloor}{3} + \frac{1}{2}$$

and

$$b_n = \frac{b_{n-1}}{3} + \frac{1}{6}$$

for $n \geq 2$. Prove that $a_n > b_n$ for all $n \in \mathbb{N}$.

Definition - Floor Function: For $x \in \mathbb{R}$, the floor function $\lfloor x \rfloor$ is the largest integer not exceeding x. For example, $\lfloor 2 \rfloor = 2$, $\lfloor 3.6 \rfloor = 3$, $\lfloor -1.2 \rfloor = -2$, etc. Formally, for any $\ell \in \mathbb{Z}$,

$$\lfloor x \rfloor = \ell \Leftrightarrow \ell \leq x < \ell + 1.$$

With this definition of floor function, it should be easy to show that $\lfloor x \rfloor$ has the property

$$\lfloor x + a \rfloor = \lfloor x \rfloor + a$$

for any $a \in \mathbb{Z}$.

Solution

Base Cases

We need to prove that it is true for $n = 1$. Since we are given that $a_1 > b_1$, it is true for $n = 1$.

Induction Step

Assume

$$a_k > b_k \tag{40.1}$$

for some $k \geq 1$. To prove the induction step, we need to prove $a_{k+1} > b_{k+1}$. By the recursive formula of a_n,

$$a_{k+1} = \frac{\lfloor a_k \rfloor}{3} + \frac{1}{2}.$$

From the definition of floor function, we know that

$$a_k < \lfloor a_k \rfloor + 1 \Leftrightarrow \lfloor a_k \rfloor > a_k - 1.$$

So,

$$a_{k+1} > \frac{a_k - 1}{3} + \frac{1}{2} = \frac{a_k}{3} + \frac{1}{6}.$$

Using assumption (40.1), since $a_k > b_k$,

$$a_{k+1} > \frac{b_k}{3} + \frac{1}{6}.$$

By the recursive formula of b_n,

$$\frac{b_k}{3} + \frac{1}{6} = b_{k+1}.$$

Hence, $a_{k+1} > b_{k+1}$. Thus, we proved that it is true for $n = k + 1$ if it is true for $n = k$.

Conclusion

We proved that it is true for $n = 1$. Then, we proved that it is true for $n = k + 1$ if it is true for $n = k$. Therefore, by weak induction,

$$a_n > b_n$$

for all $n \in \mathbb{N}$.

Problem 41

Topic: Harmonic Number, Inequality

Problem: Prove
$$H_n H_n^{(2)} + 2 \geq H_n + H_n^{(2)} + H_n^{(3)}$$
for all $n \in \mathbb{N}$ where H_n denotes the n-th Harmonic number and $H_n^{(m)}$ denotes the n-th generalized Harmonic number of order m.

Definition - Harmonic Number:
$$H_n = \sum_{j=1}^{n} \frac{1}{j}.$$

Definition - Generalized Harmonic Number:
$$H_n^{(m)} = \sum_{j=1}^{n} \frac{1}{j^m}.$$

Solution

Base Cases

We need to prove that it is true for $n = 1$. We can check that
$$H_1 H_1^{(2)} + 2 = 1 \cdot 1 + 2 = 3 \geq 3 = 1 + 1 + 1 = H_1 + H_1^{(2)} + H_1^{(3)},$$
so it is true for $n = 1$.

Induction Step

Assume

$$H_k H_k^{(2)} + 2 \geq H_k + H_k^{(2)} + H_k^{(3)} \qquad (41.1)$$

for some $k \geq 1$. To prove the induction step, we need to prove

$$H_{k+1} H_{k+1}^{(2)} + 2 \geq H_{k+1} + H_{k+1}^{(2)} + H_{k+1}^{(3)}.$$

Note that

$$H_{k+1} = \sum_{j=1}^{k+1} \frac{1}{j} = \sum_{j=1}^{k} \frac{1}{j} + \frac{1}{k+1} = H_k + \frac{1}{k+1},$$

$$H_{k+1}^{(2)} = \sum_{j=1}^{k+1} \frac{1}{j^2} = \sum_{j=1}^{k} \frac{1}{j^2} + \frac{1}{(k+1)^2} = H_k^{(2)} + \frac{1}{(k+1)^2},$$

and

$$H_{k+1}^{(3)} = \sum_{j=1}^{k+1} \frac{1}{j^3} = \sum_{j=1}^{k} \frac{1}{j^3} + \frac{1}{(k+1)^3} = H_k^{(3)} + \frac{1}{(k+1)^3}.$$

Using assumption (41.1),

$$
\begin{aligned}
H_{k+1} H_{k+1}^{(2)} + 2 &= \left(H_k + \frac{1}{k+1} \right) \left(H_k^{(2)} + \frac{1}{(k+1)^2} \right) + 2 \\
&= H_k H_k^{(2)} + \frac{H_k^{(2)}}{k+1} + \frac{H_k}{(k+1)^2} + \frac{1}{(k+1)^3} + 2 \\
&\geq H_k + H_k^{(2)} + H_k^{(3)} + + \frac{H_k^{(2)}}{k+1} + \frac{H_k}{(k+1)^2} + \frac{1}{(k+1)^3} \\
&= H_k + \frac{H_k^{(2)}}{k+1} + H_k^{(2)} + \frac{H_k}{(k+1)^2} + H_{k+1}^{(3)}.
\end{aligned}
$$

For $k \geq 1$,

$$H_k = 1 + \frac{1}{2} + \cdots + \frac{1}{k} \geq 1,$$

and

$$H_k^{(2)} = 1 + \frac{1}{2^2} + \cdots + \frac{1}{k^2} \geq 1.$$

So,

$$
\begin{aligned}
H_{k+1} H_{k+1}^{(2)} + 2 &\geq H_k + \frac{1}{k+1} + H_k^{(2)} + \frac{1}{(k+1)^2} + H_{k+1}^{(3)} \\
&= H_{k+1} + H_{k+1}^{(2)} + H_{k+1}^{(3)}.
\end{aligned}
$$

Thus, we proved that it is true for $n = k + 1$ if it is true for $n = k$.

Conclusion

We proved that it is true for $n = 1$. Then, we proved that it is true for $n = k + 1$ if it is true for $n = k$. Therefore, by weak induction,

$$H_n H_n^{(2)} + 2 \geq H_n + H_n^{(2)} + H_n^{(3)}$$

for all $n \in \mathbb{N}$.

Problem 42

Topic: Harmonic Number, Inequality

Problem: Prove

$$H_n H_n^{(2)} < \sum_{j=1}^{n} \sum_{i=1}^{j} \frac{i+j}{i^2 j^2}$$

for all $n \in \mathbb{N}$ where H_n denotes the n-th Harmonic number and $H_n^{(m)}$ denotes the n-th generalized Harmonic number of order m.

Definition - Harmonic Number:

$$H_n = \sum_{j=1}^{n} \frac{1}{j}.$$

Definition - Generalized Harmonic Number:

$$H_n^{(m)} = \sum_{j=1}^{n} \frac{1}{j^m}.$$

Solution

Base Cases

We need to prove that it is true for $n = 1$. We have

$$H_1 H_1^{(2)} = 1 \cdot 1 = 1$$

and

$$\sum_{j=1}^{1}\sum_{i=1}^{j}\frac{i+j}{i^2 j^2} = \sum_{i=1}^{1}\frac{i+1}{i^2 1^2} = \frac{1+1}{1^2 1^2} = 2.$$

So,

$$H_1 H_1^{(2)} < \sum_{j=1}^{1}\sum_{i=1}^{j}\frac{i+j}{i^2 j^2},$$

and it is true for $n = 1$.

Induction Step

Assume

$$H_k H_k^{(2)} < \sum_{j=1}^{k}\sum_{i=1}^{j}\frac{i+j}{i^2 j^2} \tag{42.1}$$

for some $k \geq 1$. To prove the induction step, we need to prove

$$H_{k+1} H_{k+1}^{(2)} < \sum_{j=1}^{k+1}\sum_{i=1}^{j}\frac{i+j}{i^2 j^2}.$$

Note that

$$\sum_{j=1}^{k+1}\sum_{i=1}^{j}\frac{i+j}{i^2 j^2} = \sum_{j=1}^{k}\sum_{i=1}^{j}\frac{i+j}{i^2 j^2} + \sum_{i=1}^{k+1}\frac{i+(k+1)}{i^2(k+1)^2}$$

$$= \sum_{j=1}^{k}\sum_{i=1}^{j}\frac{i+j}{i^2 j^2} + \frac{1}{(k+1)^2}\sum_{i=1}^{k+1}\frac{1}{i} + \frac{1}{k+1}\sum_{i=1}^{k+1}\frac{1}{i^2}$$

$$= \sum_{j=1}^{k}\sum_{i=1}^{j}\frac{i+j}{i^2 j^2} + \frac{H_{k+1}}{(k+1)^2} + \frac{H_{k+1}^{(2)}}{k+1}.$$

Using assumption (42.1),

$$\sum_{j=1}^{k+1}\sum_{i=1}^{j}\frac{i+j}{i^2 j^2} > H_k H_k^{(2)} + \frac{H_{k+1}}{(k+1)^2} + \frac{H_{k+1}^{(2)}}{k+1}$$

$$= H_k H_k^{(2)} + \frac{1}{(k+1)^2}\left(H_k + \frac{1}{k+1}\right) + \frac{1}{k+1}\left(H_k^{(2)} + \frac{1}{(k+1)^2}\right)$$

$$= H_k H_k^{(2)} + \frac{H_k}{(k+1)^2} + \frac{H_k^{(2)}}{k+1} + \frac{2}{(k+1)^3}$$

$$> H_k H_k^{(2)} + \frac{H_k}{(k+1)^2} + \frac{H_k^{(2)}}{k+1} + \frac{1}{(k+1)^3}$$

$$= \left(H_k + \frac{1}{k+1}\right)\left(H_k^{(2)} + \frac{1}{(k+1)^2}\right)$$

$$= H_{k+1} H_{k+1}^{(2)}.$$

Thus, we proved that it is true for $n = k+1$ if it is true for $n = k$.

Conclusion

We proved that it is true for $n = 1$. Then, we proved that it is true for $n = k + 1$ if it is true for $n = k$. Therefore, by weak induction,

$$H_n H_n^{(2)} < \sum_{j=1}^{n} \sum_{i=1}^{j} \frac{i + j}{i^2 j^2}$$

for all $n \in \mathbb{N}$.

Problem 43

Topic: Harmonic Number, Inequality

Problem: Prove
$$H_{2^n} \geq 1 + \frac{n}{2}$$
for all $n \in \mathbb{N}$ where H_n denotes the n-th Harmonic number.

Definition - Harmonic Number:
$$H_n = \sum_{j=1}^{n} \frac{1}{j}.$$

Solution

Base Cases

We need to prove that it is true for $n = 1$. We can check that
$$H_{2^1} = 1 + \frac{1}{2} \geq 1 + \frac{1}{2},$$
so it is true for $n = 1$.

Induction Step

Assume
$$H_{2^k} \geq 1 + \frac{k}{2} \tag{43.1}$$
for some $k \geq 1$. To prove the induction step, we need to prove
$$H_{2^{k+1}} \geq 1 + \frac{k+1}{2}.$$

Note that

$$H_{2^{k+1}} = \sum_{j=1}^{2^{k+1}} \frac{1}{j} = \sum_{j=1}^{2^k + 2^k} \frac{1}{j}$$

$$= \frac{1}{1} + \frac{1}{2} + \cdots + \frac{1}{2^k} + \frac{1}{2^k + 1} + \frac{1}{2^k + 2} + \cdots + \frac{1}{2^k + 2^k}$$

$$= H_{2^k} + \frac{1}{2^k + 1} + \frac{1}{2^k + 2} + \cdots + \frac{1}{2^k + 2^k}.$$

By assumption (43.1),

$$H_{2^{k+1}} \geq 1 + \frac{k}{2} + \frac{1}{2^k + 1} + \frac{1}{2^k + 2} + \cdots + \frac{1}{2^k + 2^k}.$$

Since $2^k + 1 \leq 2^k + 2 \leq \cdots \leq 2^k + 2^k$,

$$\frac{1}{2^k + 1} \geq \frac{1}{2^k + 2} \geq \cdots \geq \frac{1}{2^k + 2^k}.$$

So,

$$H_{2^{k+1}} \geq 1 + \frac{k}{2} + \underbrace{\frac{1}{2^k + 2^k} + \frac{1}{2^k + 2^k} + \cdots + \frac{1}{2^k + 2^k}}_{2^k \text{ terms}}$$

$$= 1 + \frac{k}{2} + \frac{2^k}{2^k + 2^k}$$

$$= 1 + \frac{k}{2} + \frac{1}{2}$$

$$= 1 + \frac{k+1}{2}.$$

Thus, we proved that it is true for $n = k + 1$ if it is true for $n = k$.

Conclusion

We proved that it is true for $n = 1$. Then, we proved that it is true for $n = k + 1$ if it is true for $n = k$. Therefore, by weak induction,

$$H_{2^n} \geq 1 + \frac{n}{2}$$

for all $n \in \mathbb{N}$.

Problem 44

Topic: Series, Trigonometry

Problem: Prove

$$\sum_{j=1}^{n} \cos j = \frac{\cos\left(\frac{n+1}{2}\right)\sin\left(\frac{n}{2}\right)}{\sin\left(\frac{1}{2}\right)}.$$

Solution

Base Cases

We need to prove that it is true for $n = 1$. We can check that

$$\sum_{j=1}^{1} \cos j = \cos 1 = \frac{\cos\left(\frac{1+1}{2}\right)\sin\left(\frac{1}{2}\right)}{\sin\left(\frac{1}{2}\right)},$$

so it is true for $n = 1$.

Induction Step

Assume

$$\sum_{j=1}^{k} \cos j = \frac{\cos\left(\frac{k+1}{2}\right)\sin\left(\frac{k}{2}\right)}{\sin\left(\frac{1}{2}\right)} \tag{44.1}$$

for some $k \geq 1$. To prove the induction step, we need to prove

$$\sum_{j=1}^{k+1} \cos j = \frac{\cos\left(\frac{k+2}{2}\right)\sin\left(\frac{k+1}{2}\right)}{\sin\left(\frac{1}{2}\right)}.$$

Using assumption (44.1),

$$\sum_{j=1}^{k+1} \cos j = \sum_{j=1}^{k} \cos j + \cos(k+1)$$

$$= \frac{\cos\left(\frac{k+1}{2}\right)\sin\left(\frac{k}{2}\right)}{\sin\left(\frac{1}{2}\right)} + \cos(k+1).$$

Multiplying both sides by $\sin\left(\frac{1}{2}\right)$,

$$\sin\left(\frac{1}{2}\right)\sum_{j=1}^{k+1}\cos j = \cos\left(\frac{k+1}{2}\right)\sin\left(\frac{k}{2}\right) + \cos(k+1)\sin\left(\frac{1}{2}\right).$$

Using the formula

$$\cos(a)\sin(b) = \frac{1}{2}[\sin(a+b) - \sin(a-b)],$$

we have

$$\cos\left(\frac{k+1}{2}\right)\sin\left(\frac{k}{2}\right) = \frac{1}{2}\left[\sin\left(k+\frac{1}{2}\right) - \sin\left(\frac{1}{2}\right)\right]$$

and

$$\cos(k+1)\sin\left(\frac{1}{2}\right) = \frac{1}{2}\left[\sin\left(k+\frac{3}{2}\right) - \sin\left(k+\frac{1}{2}\right)\right].$$

Hence,

$$\cos\left(\frac{k+1}{2}\right)\sin\left(\frac{k}{2}\right) + \cos(k+1)\sin\left(\frac{1}{2}\right) = \frac{1}{2}\left[\sin\left(k+\frac{3}{2}\right) - \sin\left(\frac{1}{2}\right)\right]$$

$$= \cos\left(\frac{k+2}{2}\right)\sin\left(\frac{k+1}{2}\right),$$

and we have

$$\sin\left(\frac{1}{2}\right)\sum_{j=1}^{k+1}\cos j = \cos\left(\frac{k+2}{2}\right)\sin\left(\frac{k+1}{2}\right)$$

$$\sum_{j=1}^{k+1}\cos j = \frac{\cos\left(\frac{k+2}{2}\right)\sin\left(\frac{k+1}{2}\right)}{\sin\left(\frac{1}{2}\right)}.$$

Thus, we proved that it is true for $n = k+1$ if it is true for $n = k$.

Conclusion

We proved that it is true for $n = 1$. Then, we proved that it is true for $n = k+1$ if it is true for $n = k$. Therefore, by weak induction,

$$\sum_{j=1}^{n}\cos j = \frac{\cos\left(\frac{n+1}{2}\right)\sin\left(\frac{n}{2}\right)}{\sin\left(\frac{1}{2}\right)}$$

for all $n \in \mathbb{N}$.

Problem 45

Topic: Series, Trigonometry

Problem: Prove

$$\sum_{j=1}^{n} \sin j = \frac{\sin\left(\frac{n+1}{2}\right)\sin\left(\frac{n}{2}\right)}{\sin\left(\frac{1}{2}\right)}.$$

Solution

Base Cases

We need to prove that it is true for $n = 1$. We can check that

$$\sum_{j=1}^{1} \sin j = \sin 1 = \frac{\sin\left(\frac{1+1}{2}\right)\sin\left(\frac{1}{2}\right)}{\sin\left(\frac{1}{2}\right)},$$

so it is true for $n = 1$.

Induction Step

Assume

$$\sum_{j=1}^{k} \sin j = \frac{\sin\left(\frac{k+1}{2}\right)\sin\left(\frac{k}{2}\right)}{\sin\left(\frac{1}{2}\right)} \tag{45.1}$$

for some $k \geq 1$. To prove the induction step, we need to prove

$$\sum_{j=1}^{k+1} \sin j = \frac{\sin\left(\frac{k+2}{2}\right)\sin\left(\frac{k+1}{2}\right)}{\sin\left(\frac{1}{2}\right)}.$$

Using assumption (45.1),

$$\sum_{j=1}^{k+1} \sin j = \sum_{j=1}^{k} \sin j + \sin (k+1)$$

$$= \frac{\sin \left(\frac{k+1}{2}\right) \sin \left(\frac{k}{2}\right)}{\sin \left(\frac{1}{2}\right)} + \sin (k+1).$$

Multiplying both sides by $\sin \left(\frac{1}{2}\right)$,

$$\sin \left(\frac{1}{2}\right) \sum_{j=1}^{k+1} \sin j = \sin \left(\frac{k+1}{2}\right) \sin \left(\frac{k}{2}\right) + \sin (k+1) \sin \left(\frac{1}{2}\right).$$

Using the formula

$$\sin (a) \sin (b) = \frac{1}{2} [\cos (a-b) - \cos (a+b)],$$

we have

$$\sin \left(\frac{k+1}{2}\right) \sin \left(\frac{k}{2}\right) = \frac{1}{2} \left[\cos \left(\frac{1}{2}\right) - \cos \left(k+\frac{1}{2}\right)\right]$$

and

$$\sin (k+1) \sin \left(\frac{1}{2}\right) = \frac{1}{2} \left[\cos \left(k+\frac{1}{2}\right) - \cos \left(k+\frac{3}{2}\right)\right].$$

Hence,

$$\sin \left(\frac{k+1}{2}\right) \sin \left(\frac{k}{2}\right) + \sin (k+1) \sin \left(\frac{1}{2}\right) = \frac{1}{2} \left[\cos \left(\frac{1}{2}\right) - \cos \left(k+\frac{3}{2}\right)\right]$$

$$= \sin \left(\frac{k+2}{2}\right) \sin \left(\frac{k+1}{2}\right),$$

and we have

$$\sin \left(\frac{1}{2}\right) \sum_{j=1}^{k+1} \sin j = \sin \left(\frac{k+2}{2}\right) \sin \left(\frac{k+1}{2}\right)$$

$$\sum_{j=1}^{k+1} \sin j = \frac{\sin \left(\frac{k+2}{2}\right) \sin \left(\frac{k+1}{2}\right)}{\sin \left(\frac{1}{2}\right)}.$$

Thus, we proved that it is true for $n = k+1$ if it is true for $n = k$.

Conclusion

We proved that it is true for $n = 1$. Then, we proved that it is true for $n = k+1$ if it is true for $n = k$. Therefore, by weak induction,

$$\sum_{j=1}^{n} \sin j = \frac{\sin \left(\frac{n+1}{2}\right) \sin \left(\frac{n}{2}\right)}{\sin \left(\frac{1}{2}\right)}$$

for all $n \in \mathbb{N}$.

Problem 46

Topic: Fibonacci Number, Golden Number

Problem: Prove
$$\varphi^n = F_n\varphi + F_{n-1}$$
for all $n \in \mathbb{N}$ where φ is the golden number and F_n denotes the n-th Fibonacci number.

Definition - Golden Number: The golden number φ is the positive real solution of $x^2 - x - 1 = 0$. Its exact value is
$$\varphi = \frac{1 + \sqrt{5}}{2}.$$

Definition - Fibonacci Numbers: The Fibonacci numbers are a sequence of numbers defined as $F_0 = 0$, $F_1 = 1$, and $F_n = F_{n-1} + F_{n-2}$ for $n \geq 2$.

Solution

Base Cases

We need to prove that it is true for $n = 1$. We can check that
$$F_1\varphi + F_0 = \varphi = \varphi^1$$
because $F_1 = 1$ and $F_0 = 0$, so it is true for $n = 1$.

Induction Step

Assume

$$\varphi^k = F_k\varphi + F_{k-1} \qquad\qquad (46.1)$$

for some $k \geq 1$. To prove the induction step, we need to prove

$$\varphi^{k+1} = F_{k+1}\varphi + F_k.$$

Multiplying both sides of (46.1) by φ,

$$\varphi^{k+1} = F_k\varphi^2 + F_{k-1}\varphi.$$

From the definition of φ, we have

$$\varphi^2 - \varphi - 1 = 0 \Leftrightarrow \varphi^2 = \varphi + 1.$$

So,

$$\begin{aligned}
\varphi^{k+1} &= F_k(\varphi + 1) + F_{k-1}\varphi \\
&= (F_{k-1} + F_k)\varphi + F_k.
\end{aligned}$$

Using the recursive relation of F_n,

$$F_{k-1} + F_k = F_{k+1}.$$

Hence,

$$\varphi^{k+1} = F_{k+1}\varphi + F_k.$$

Thus, we proved that it is true for $n = k + 1$ if it is true for $n = k$.

Conclusion

We proved that it is true for $n = 1$. Then, we proved that it is true for $n = k + 1$ if it is true for $n = k$. Therefore, by weak induction,

$$\varphi^n = F_n\varphi + F_{n-1}$$

for all $n \in \mathbb{N}$.

Problem 47

Topic: Fibonacci Number, Golden Number

Problem: Prove Binet's formula for F_n,

$$F_n = \frac{\varphi^n - (1 - \varphi)^n}{\sqrt{5}},$$

where F_n denotes the n-th Fibonacci number and φ is the golden number.

Definition - Golden Number: The golden number φ is the positive real solution of $x^2 - x - 1 = 0$. Its exact value is

$$\varphi = \frac{1 + \sqrt{5}}{2}.$$

Definition - Fibonacci Numbers: The Fibonacci numbers are a sequence of numbers defined as $F_0 = 0$, $F_1 = 1$, and $F_n = F_{n-1} + F_{n-2}$ for $n \geq 2$.

Solution

Base Cases

We need to prove that it is true for $n = 1$ and $n = 2$. When $n = 1$,

$$\frac{\varphi^1 - (1 - \varphi)^1}{\sqrt{5}} = \frac{2\varphi - 1}{\sqrt{5}} = \frac{2 \cdot \left(\frac{1+\sqrt{5}}{2}\right) - 1}{\sqrt{5}} = 1 = F_1.$$

When $n = 2$,

$$\frac{\varphi^2 - (1-\varphi)^2}{\sqrt{5}} = \frac{\varphi^2 - (1 - 2\varphi + \varphi^2)}{\sqrt{5}} = \frac{2\varphi - 1}{\sqrt{5}} = 1 = 1 + 0 = F_1 + F_0 = F_2.$$

So, it is true for $n = 1$ and $n = 2$.

Induction Step

Assume

$$F_m = \frac{\varphi^m - (1-\varphi)^m}{\sqrt{5}} \tag{47.1}$$

for all $1 \leq m \leq k$ for some $k \geq 2$. To prove the induction step, we need to prove

$$F_{k+1} = \frac{\varphi^{k+1} - (1-\varphi)^{k+1}}{\sqrt{5}}.$$

Using the recursive relation of F_n,

$$F_{k+1} = F_{k-1} + F_k.$$

By assumption (47.1) when $m = k - 1$,

$$F_{k-1} = \frac{\varphi^{k-1} - (1-\varphi)^{k-1}}{\sqrt{5}}.$$

By assumption (47.1) when $m = k$,

$$F_k = \frac{\varphi^k - (1-\varphi)^k}{\sqrt{5}}.$$

So,

$$\begin{aligned}
F_{k+1} &= \frac{\varphi^{k-1} - (1-\varphi)^{k-1}}{\sqrt{5}} + \frac{\varphi^k - (1-\varphi)^k}{\sqrt{5}} \\
&= \frac{\varphi^{k-1}(1 + \varphi) - (1-\varphi)^{k-1}(1 + 1 - \varphi)}{\sqrt{5}}.
\end{aligned}$$

From the definition of φ, we have

$$\varphi^2 - \varphi - 1 = 0 \Leftrightarrow \varphi^2 = \varphi + 1.$$

We can check that $1 - \varphi$ is also a solution of $x^2 - x - 1 = 0$ since

$$\begin{aligned}
(1-\varphi)^2 - (1-\varphi) - 1 &= 1 - 2\varphi + \varphi^2 - 1 + \varphi - 1 \\
&= \varphi^2 - \varphi - 1 = 0.
\end{aligned}$$

So,

$$(1-\varphi)^2 = 1 - \varphi + 1.$$

Hence,

$$F_{k+1} = \frac{\varphi^{k-1} \cdot \varphi^2 - (1-\varphi)^{k-1} \cdot (1-\varphi)^2}{\sqrt{5}}$$
$$= \frac{\varphi^{k+1} - (1-\varphi)^{k+1}}{\sqrt{5}}.$$

Thus, we proved that it is true for $n = k+1$ if it is true for $n = m$ for all $1 \leq m \leq k$.

Conclusion

We proved that it is true for $n = 1$ and $n = 2$. Then, we proved that it is true for $n = k + 1$ if it is true for $n = m$ for all $1 \leq m \leq k$. Therefore, by strong induction,

$$F_n = \frac{\varphi^n - (1-\varphi)^n}{\sqrt{5}}$$

for all $n \in \mathbb{N}$.

Problem 48

Topic: Lucas Number, Golden Number

Problem: Prove Binet's formula for L_n,

$$L_n = \varphi^n + (1 - \varphi)^n,$$

where L_n denotes the n-th Lucas number and φ is the golden number.

Definition - Golden Number: The golden number φ is the positive real solution of $x^2 - x - 1 = 0$. Its exact value is

$$\varphi = \frac{1 + \sqrt{5}}{2}.$$

Definition - Lucas Numbers: The Lucas numbers are a sequence of numbers defined as $L_0 = 2$, $L_1 = 1$, and $L_n = L_{n-1} + L_{n-2}$ for $n \geq 2$.

Solution

Base Cases

We need to prove that it is true for $n = 1$ and $n = 2$. When $n = 1$,

$$\varphi^1 + (1 - \varphi)^1 = 1 = L_1.$$

When $n = 2$,

$$\varphi^2 + (1 - \varphi)^2 = \varphi^2 + (1 - 2\varphi + \varphi^2) = 2\varphi^2 - 2\varphi + 1.$$

From the definition of φ, we have

$$\varphi^2 - \varphi - 1 = 0 \Leftrightarrow \varphi^2 = \varphi + 1,$$

so

$$\varphi^2 + (1 - \varphi)^2 = 2(\varphi + 1) - 2\varphi + 1 = 3 = 2 + 1 = L_0 + L_1 = L_2.$$

So, it is true for $n = 1$ and $n = 2$.

Induction Step

Assume

$$L_m = \varphi^m + (1 - \varphi)^m \tag{48.1}$$

for all $1 \leq m \leq k$ for some $k \geq 2$. To prove the induction step, we need to prove

$$L_{k+1} = \varphi^{k+1} + (1 - \varphi)^{k+1}.$$

Using the recursive relation of L_n,

$$L_{k+1} = L_{k-1} + L_k.$$

By assumption (48.1) when $m = k - 1$,

$$L_{k-1} = \varphi^{k-1} + (1 - \varphi)^{k-1}.$$

By assumption (48.1) when $m = k$,

$$L_k = \varphi^k + (1 - \varphi)^k.$$

So,

$$\begin{aligned}
L_{k+1} &= \varphi^{k-1} + (1 - \varphi)^{k-1} + \varphi^k + (1 - \varphi)^k \\
&= \varphi^{k-1}(1 + \varphi) + (1 - \varphi)^{k-1}(1 + 1 - \varphi).
\end{aligned}$$

From the definition of φ, we have

$$\varphi^2 - \varphi - 1 = 0 \Leftrightarrow \varphi^2 = \varphi + 1.$$

We can check that $1 - \varphi$ is also a solution of $x^2 - x - 1 = 0$ since

$$\begin{aligned}
(1 - \varphi)^2 - (1 - \varphi) - 1 &= 1 - 2\varphi + \varphi^2 - 1 + \varphi - 1 \\
&= \varphi^2 - \varphi - 1 = 0.
\end{aligned}$$

So,

$$(1 - \varphi)^2 = 1 - \varphi + 1.$$

Hence,

$$\begin{aligned}
L_{k+1} &= \varphi^{k-1} \cdot \varphi^2 + (1 - \varphi)^{k-1} \cdot (1 - \varphi)^2 \\
&= \varphi^{k+1} + (1 - \varphi)^{k+1}.
\end{aligned}$$

Thus, we proved that it is true for $n = k+1$ if it is true for $n = m$ for all $1 \leq m \leq k$.

Conclusion

We proved that it is true for $n = 1$ and $n = 2$. Then, we proved that it is true for $n = k + 1$ if it is true for $n = m$ for all $1 \leq m \leq k$. Therefore, by strong induction,

$$L_n = \varphi^n + (1 - \varphi)^n$$

for all $n \in \mathbb{N}$.

Problem 49

Topic: Fibonacci Number, Lucas Number

Problem: Prove
$$L_n = F_{n-1} + F_{n+1}$$
where L_n denotes the n-th Lucas number and F_n denotes the n-th Fibonacci number.

Definition - Lucas Numbers: The Lucas numbers are a sequence of numbers defined as $L_0 = 2$, $L_1 = 1$, and $L_n = L_{n-1} + L_{n-2}$ for $n \geq 2$.

Definition - Fibonacci Numbers: The Fibonacci numbers are a sequence of numbers defined as $F_0 = 0$, $F_1 = 1$, and $F_n = F_{n-1} + F_{n-2}$ for $n \geq 2$.

Solution

Base Cases

We need to prove that it is true for $n = 1$ and $n = 2$. When $n = 1$,
$$F_0 + F_2 = F_0 + F_0 + F_1 = 0 + 0 + 1 = 1 = L_1.$$

When $n = 2$,
$$F_1 + F_3 = F_1 + F_1 + F_2 = F_1 + F_1 + F_0 + F_1 = 1 + 1 + 0 + 1 = 3$$

and
$$L_2 = L_0 + L_1 = 2 + 1 = 3,$$

so
$$F_1 + F_3 = L_2.$$

So, it is true for $n = 1$ and $n = 2$.

Induction Step

Assume

$$L_m = F_{m-1} + F_{m+1} \tag{49.1}$$

for all $1 \leq m \leq k$ for some $k \geq 2$. To prove the induction step, we need to prove

$$L_{k+1} = F_k + F_{k+2}.$$

Using the recursive relation of L_n,

$$L_{k+1} = L_{k-1} + L_k.$$

By assumption (49.1) when $m = k - 1$,

$$L_{k-1} = F_{k-2} + F_k.$$

By assumption (49.1) when $m = k$,

$$L_k = F_{k-1} + F_{k+1}.$$

So,

$$L_{k+1} = F_{k-2} + F_k + F_{k-1} + F_{k+1}$$
$$= (F_{k-2} + F_{k-1}) + (F_k + F_{k+1}).$$

Using the recursive relation of F_n,

$$F_{k-2} + F_{k-1} = F_k,$$

and

$$F_k + F_{k+1} = F_{k+2}.$$

Hence,

$$L_{k+1} = F_k + F_{k+2}.$$

Thus, we proved that it is true for $n = k+1$ if it is true for $n = m$ for all $1 \leq m \leq k$.

Conclusion

We proved that it is true for $n = 1$ and $n = 2$. Then, we proved that it is true for $n = k + 1$ if it is true for $n = m$ for all $1 \leq m \leq k$. Therefore, by strong induction,

$$L_n = F_{n-1} + F_{n+1}$$

for all $n \in \mathbb{N}$.

Problem 50

Topic: Sequence, Fibonacci Number

Problem: Let a_1 and a_2 be arbitrary real numbers, and

$$a_n = a_{n-1} + a_{n-2}$$

for $n \geq 3$. Prove that

$$a_n = a_1 F_{n-2} + a_2 F_{n-1}$$

for $n \geq 2$ where F_n denotes the n-th Fibonacci number.

Definition - Fibonacci Numbers: The Fibonacci numbers are a sequence of numbers defined as $F_0 = 0$, $F_1 = 1$, and $F_n = F_{n-1} + F_{n-2}$ for $n \geq 2$.

Solution

Base Cases

We need to prove that it is true for $n = 2$ and $n = 3$. When $n = 2$,

$$a_2 = a_1 F_0 + a_2 F_1$$

because $F_0 = 0$ and $F_1 = 1$. When $n = 3$, by the recursive formula of a_n,

$$a_3 = a_1 + a_2 = a_1 F_1 + a_2 F_2$$

because $F_1 = 1$ and $F_2 = F_0 + F_1 = 0 + 1 = 1$. So, it is true for $n = 2$ and $n = 3$.

Induction Step

Assume

$$a_m = a_1 F_{m-2} + a_2 F_{m-1} \tag{50.1}$$

for all $2 \leq m \leq k$ for some $k \geq 3$. To prove the induction step, we need to prove

$$a_{k+1} = a_1 F_{k-1} + a_2 F_k.$$

By the recursive formula of a_n,

$$a_{k+1} = a_{k-1} + a_k.$$

By assumption (50.1) when $m = k - 1$,

$$a_{k-1} = a_1 F_{k-3} + a_2 F_{k-2}.$$

By assumption (50.1) when $m = k$,

$$a_k = a_1 F_{k-2} + a_2 F_{k-1}.$$

So,

$$\begin{aligned}
a_{k+1} &= a_1 F_{k-3} + a_2 F_{k-2} + a_1 F_{k-2} + a_2 F_{k-1} \\
&= a_1 (F_{k-3} + F_{k-2}) + a_2 (F_{k-2} + F_{k-1}).
\end{aligned}$$

Using the recursive relation of F_n,

$$F_{k-3} + F_{k-2} = F_{k-1},$$

and

$$F_{k-2} + F_{k-1} = F_k.$$

Hence,

$$a_{k+1} = a_1 F_{k-1} + a_2 F_k.$$

Thus, we proved that it is true for $n = k+1$ if it is true for $n = m$ for all $2 \leq m \leq k$.

Conclusion

We proved that it is true for $n = 2$ and $n = 3$. Then, we proved that it is true for $n = k + 1$ if it is true for $n = m$ for all $2 \leq m \leq k$. Therefore, by strong induction,

$$a_n = a_1 F_{n-2} + a_2 F_{n-1}$$

for $n \geq 2$.

$$f^n(x) = (f \circ f \circ \cdots \circ f)(x)$$

Problem 51

Topic: Fibonacci Number

Problem: For $\ell, n \in \mathbb{N}$, prove

$$F_{\ell+n} = F_{\ell-1}F_n + F_\ell F_{n+1}$$

where F_n denotes the n-th Fibonacci number.

Definition - Fibonacci Numbers: The Fibonacci numbers are a sequence of numbers defined as $F_0 = 0$, $F_1 = 1$, and $F_n = F_{n-1} + F_{n-2}$ for $n \geq 2$.

Solution

Base Cases

We need to prove that it is true for $n = 1$ and $n = 2$. When $n = 1$, by the recursive relation of F_n,

$$F_{\ell+1} = F_{\ell-1} + F_\ell = F_{\ell-1}F_1 + F_\ell F_2$$

because $F_1 = 1$ and $F_2 = F_0 + F_1 = 0 + 1 = 1$. When $n = 2$, by the recursive relation of F_n,

$$F_{\ell+2} = F_\ell + F_{\ell+1} = F_\ell + F_{\ell-1} + F_\ell = F_{\ell-1} + 2F_\ell = F_{\ell-1}F_2 + F_\ell F_3$$

because $F_2 = 1$ and $F_3 = F_1 + F_2 = 1 + 1 = 2$. So, it is true for $n = 1$ and $n = 2$.

Induction Step

Assume

$$F_{\ell+m} = F_{\ell-1}F_m + F_\ell F_{m+1} \tag{51.1}$$

for all $1 \le m \le k$ for some $k \ge 2$. To prove the induction step, we need to prove

$$F_{\ell+k+1} = F_{\ell-1}F_{k+1} + F_\ell F_{k+2}.$$

Using the recursive relation of F_n,

$$F_{\ell+k+1} = F_{\ell+k-1} + F_{\ell+k}.$$

By assumption (51.1) when $m = k - 1$,

$$F_{\ell+k-1} = F_{\ell-1}F_{k-1} + F_\ell F_k.$$

By assumption (51.1) when $m = k$,

$$F_{\ell+k} = F_{\ell-1}F_k + F_\ell F_{k+1}.$$

So,

$$\begin{aligned}
F_{\ell+k+1} &= F_{\ell-1}F_{k-1} + F_\ell F_k + F_{\ell-1}F_k + F_\ell F_{k+1} \\
&= F_{\ell-1}(F_{k-1} + F_k) + F_\ell(F_k + F_{k+1}).
\end{aligned}$$

Using the recursive relation of F_n,

$$F_{k-1} + F_k = F_{k+1},$$

and

$$F_k + F_{k+1} = F_{k+2}.$$

Hence,

$$F_{\ell+k+1} = F_{\ell-1}F_{k+1} + F_\ell F_{k+2}.$$

Thus, we proved that it is true for $n = k+1$ if it is true for $n = m$ for all $1 \le m \le k$

Conclusion

We proved that it is true for $n = 1$ and $n = 2$. Then, we proved that it is true for $n = k+1$ if it is true for $n = m$ for all $1 \le m \le k$. Therefore, by strong induction

$$F_{\ell+n} = F_{\ell-1}F_n + F_\ell F_{n+1}$$

for all $n \in \mathbb{N}$.

Problem 52

Topic: Iterated Function

Problem: For $a, b \in \mathbb{R}$ and $a \neq 0$, if $f(x) = ax + b$, then prove

$$f^n(x) = a^n x + b \sum_{j=0}^{n-1} a^j$$

for all $n \in \mathbb{N}$ where $f^n(x)$ denotes the n-th iterate of $f(x)$.

Definition - Iterated Function: The n-th iterate of $f(x)$ is the composition of $f(x)$ with itself n times, i.e.

$$f^n(x) = (\underbrace{f \circ f \circ \cdots \circ f}_{n \text{ times}})(x).$$

Solution

Base Cases

We need to prove that it is true for $n = 1$. We can check that

$$f^1(x) = a^1 x + b \sum_{j=0}^{1-1} a^j = ax + b = f(x),$$

so it is true for $n = 1$.

Induction Step

Assume

$$f^k(x) = a^k x + b \sum_{j=0}^{k-1} a^j \tag{52.1}$$

for some $k \geq 1$. To prove the induction step, we need to prove

$$f^{k+1}(x) = a^{k+1} x + b \sum_{j=0}^{k} a^j.$$

Using assumption (52.1),

$$f^{k+1}(x) = \left(f^k \circ f\right)(x) = a^k(ax + b) + b \sum_{j=0}^{k-1} a^j$$

$$= a^{k+1} x + ba^k + b \sum_{j=0}^{k-1} a^j$$

$$= a^{k+1} x + b \sum_{j=0}^{k} a^j.$$

Thus, we proved that it is true for $n = k + 1$ if it is true for $n = k$.

Conclusion

We proved that it is true for $n = 1$. Then, we proved that it is true for $n = k + 1$ if it is true for $n = k$. Therefore, by weak induction,

$$f^n(x) = a^n x + b \sum_{j=0}^{n-1} a^j$$

for all $n \in \mathbb{N}$.

Problem 53

Topic: Iterated Function, Fibonacci Number

Problem: If

$$f(x) = \frac{1}{1+x},$$

then prove

$$f^n(x) = \frac{F_{n-1}x + F_n}{F_n x + F_{n+1}}$$

for all $n \in \mathbb{N}$ where $f^n(x)$ denotes the n-th iterate of $f(x)$ and F_n denotes the n-th Fibonacci number.

Definition - Iterated Function: The n-th iterate of $f(x)$ is the composition of $f(x)$ with itself n times, i.e.

$$f^n(x) = (\underbrace{f \circ f \circ \cdots \circ f}_{n \text{ times}})(x).$$

Definition - Fibonacci Numbers: The Fibonacci numbers are a sequence of numbers defined as $F_0 = 0$, $F_1 = 1$, and $F_n = F_{n-1} + F_{n-2}$ for $n \geq 2$.

Solution

Base Cases

We need to prove that it is true for $n = 1$. We can check that

$$f^1(x) = \frac{F_0 x + F_1}{F_1 x + F_2} = \frac{0 \cdot x + 1}{1 \cdot x + 1} = \frac{1}{1+x} = f(x)$$

because $F_0 = 0$, $F_1 = 1$, and $F_2 = F_0 + F_1 = 0 + 1 = 1$, so it is true for $n = 1$.

Induction Step

Assume

$$f^k(x) = \frac{F_{k-1}x + F_k}{F_k x + F_{k+1}} \tag{53.1}$$

for some $k \geq 1$. To prove the induction step, we need to prove

$$f^{k+1}(x) = \frac{F_k x + F_{k+1}}{F_{k+1}x + F_{k+2}}.$$

Using assumption (53.1),

$$f^{k+1}(x) = \left(f^k \circ f\right)(x) = \frac{F_{k-1} \cdot \frac{1}{1+x} + F_k}{F_k \cdot \frac{1}{1+x} + F_{k+1}}$$

$$= \frac{F_{k-1} \cdot \frac{1}{1+x} + F_k}{F_k \cdot \frac{1}{1+x} + F_{k+1}} \cdot \frac{1+x}{1+x}$$

$$= \frac{F_{k-1} + F_k + F_k x}{F_k + F_{k+1} + F_{k+1}x}.$$

Using the recursive relation of F_n,

$$F_{k-1} + F_k = F_{k+1},$$

and

$$F_k + F_{k+1} = F_{k+2}.$$

Hence,

$$f^{k+1}(x) = \frac{F_k x + F_{k+1}}{F_{k+1}x + F_{k+2}}.$$

Thus, we proved that it is true for $n = k + 1$ if it is true for $n = k$.

Conclusion

We proved that it is true for $n = 1$. Then, we proved that it is true for $n = k + 1$ if it is true for $n = k$. Therefore, by weak induction,

$$f^n(x) = \frac{F_{n-1}x + F_n}{F_n x + F_{n+1}}$$

for all $n \in \mathbb{N}$.

Problem 54

Topic: Iterated Function

Problem: If
$$f(x) = \sqrt{x^2 + 1},$$
then prove
$$f^n(x) = \sqrt{x^2 + n}$$
for all $n \in \mathbb{N}$ where $f^n(x)$ denotes the n-th iterate of $f(x)$.

Definition - Iterated Function: The n-th iterate of $f(x)$ is the composition of $f(x)$ with itself n times, i.e.

$$f^n(x) = \underbrace{(f \circ f \circ \cdots \circ f)}_{n \text{ times}}(x).$$

Solution

Base Cases

We need to prove that it is true for $n = 1$. We can check that

$$f^1(x) = \sqrt{x^2 + 1} = f(x),$$

so it is true for $n = 1$.

Induction Step

Assume

$$f^k(x) = \sqrt{x^2 + k} \tag{54.1}$$

for some $k \geq 1$. To prove the induction step, we need to prove

$$f^{k+1}(x) = \sqrt{x^2 + k + 1}.$$

Using assumption (54.1),

$$f^{k+1}(x) = \left(f^k \circ f\right)(x) = \sqrt{\left(\sqrt{x^2 + 1}\right)^2 + k}$$
$$= \sqrt{x^2 + k + 1}.$$

Thus, we proved that it is true for $n = k + 1$ if it is true for $n = k$.

Conclusion

We proved that it is true for $n = 1$. Then, we proved that it is true for $n = k + 1$ if it is true for $n = k$. Therefore, by weak induction,

$$f^n(x) = \sqrt{x^2 + n}$$

for all $n \in \mathbb{N}$.

Problem 55

Topic: Iterated Function

Problem: For $a \in \mathbb{R}$ and some function g, if

$$f(x) = g^{-1}(g(x) + a),$$

then prove

$$f^n(x) = g^{-1}(g(x) + na)$$

for all $n \in \mathbb{N}$ where $f^n(x)$ denotes the n-th iterate of $f(x)$.

Definition – Iterated Function: The n-th iterate of $f(x)$ is the composition of $f(x)$ with itself n times, i.e.

$$f^n(x) = (\underbrace{f \circ f \circ \cdots \circ f}_{n \text{ times}})(x).$$

Solution

Base Cases

We need to prove that it is true for $n = 1$. We can check that

$$f^1(x) = g^{-1}(g(x) + 1 \cdot a) = g^{-1}(g(x) + a) = f(x),$$

so it is true for $n = 1$.

Induction Step

Assume

$$f^k(x) = g^{-1}(g(x) + ka) \tag{55.1}$$

for some $k \geq 1$. To prove the induction step, we need to prove

$$f^{k+1}(x) = g^{-1}(g(x) + (k+1)a).$$

Using assumption (55.1),

$$f^{k+1}(x) = \left(f^k \circ f\right)(x)$$
$$= g^{-1}(g(f(x)) + ka).$$

Since

$$g(f(x)) = g(g^{-1}(g(x) + a)) = g(x) + a,$$

we have

$$f^{k+1}(x) = g^{-1}(g(x) + a + ka)$$
$$= g^{-1}(g(x) + (k+1)a).$$

Thus, we proved that it is true for $n = k+1$ if it is true for $n = k$.

Conclusion

We proved that it is true for $n = 1$. Then, we proved that it is true for $n = k+1$ if it is true for $n = k$. Therefore, by weak induction,

$$f^n(x) = g^{-1}(g(x) + na)$$

for all $n \in \mathbb{N}$.

Problem 56

Topic: Iterated Function

Problem: If
$$f(x) = 2x(1-x),$$
then prove
$$f^n(x) = \frac{1}{2}\left[1 - (1-2x)^{2^n}\right]$$
for all $n \in \mathbb{N}$ where $f^n(x)$ denotes the n-th iterate of $f(x)$.

Definition - Iterated Function: The n-th iterate of $f(x)$ is the composition of $f(x)$ with itself n times, i.e.
$$f^n(x) = (\underbrace{f \circ f \circ \cdots \circ f}_{n \text{ times}})(x).$$

Solution

Base Cases

We need to prove that it is true for $n = 1$. We can check that
$$
\begin{aligned}
f^1(x) &= \frac{1}{2}\left[1 - (1-2x)^{2^1}\right] \\
&= \frac{1}{2}\left[4x - 4x^2\right] \\
&= 2x - 2x^2 \\
&= 2x(1-x) = f(x),
\end{aligned}
$$
so it is true for $n = 1$.

Induction Step

Assume

$$f^k(x) = \frac{1}{2}\left[1 - (1 - 2x)^{2^k}\right] \tag{56.1}$$

for some $k \geq 1$. To prove the induction step, we need to prove

$$f^{k+1}(x) = \frac{1}{2}\left[1 - (1 - 2x)^{2^{k+1}}\right].$$

Using assumption (56.1),

$$\begin{aligned}
f^{k+1}(x) = \left(f^k \circ f\right)(x) &= \frac{1}{2}\left[1 - (1 - 2 \cdot 2x(1 - x))^{2^k}\right] \\
&= \frac{1}{2}\left[1 - \left(1 - 4x + 4x^2\right)^{2^k}\right] \\
&= \frac{1}{2}\left[1 - (1 - 2x)^{2^{k+1}}\right].
\end{aligned}$$

Thus, we proved that it is true for $n = k + 1$ if it is true for $n = k$.

Conclusion

We proved that it is true for $n = 1$. Then, we proved that it is true for $n = k + 1$ if it is true for $n = k$. Therefore, by weak induction,

$$f^n(x) = \frac{1}{2}\left[1 - (1 - 2x)^{2^n}\right]$$

for all $n \in \mathbb{N}$.

Problem 57

Topic: Iterated Function

Problem: If
$$f(x) = 2x^2 - 1,$$
then prove
$$f^n(x) = \frac{1}{2}\left[\left(x + \sqrt{x^2 - 1}\right)^{2^n} + \left(x - \sqrt{x^2 - 1}\right)^{2^n}\right]$$
for all $n \in \mathbb{N}$ where $f^n(x)$ denotes the n-th iterate of $f(x)$.

Definition - Iterated Function: The n-th iterate of $f(x)$ is the composition of $f(x)$ with itself n times, i.e.
$$f^n(x) = \underbrace{(f \circ f \circ \cdots \circ f)}_{n \text{ times}}(x).$$

Solution

Base Cases

We need to prove that it is true for $n = 1$. We can check that

$$f^1(x) = \frac{1}{2}\left[\left(x + \sqrt{x^2 - 1}\right)^{2^1} + \left(x - \sqrt{x^2 - 1}\right)^{2^1}\right]$$
$$= \frac{1}{2}\left[x^2 + 2x\sqrt{x^2 - 1} + (x^2 - 1) + x^2 - 2x\sqrt{x^2 - 1} + (x^2 - 1)\right]$$
$$= 2x^2 - 1 = f(x),$$

so it is true for $n = 1$.

Induction Step

Assume

$$f^k(x) = \frac{1}{2}\left[\left(x + \sqrt{x^2 - 1}\right)^{2^k} + \left(x - \sqrt{x^2 - 1}\right)^{2^k}\right] \tag{57.1}$$

for some $k \geq 1$. To prove the induction step, we need to prove

$$f^{k+1}(x) = \frac{1}{2}\left[\left(x + \sqrt{x^2 - 1}\right)^{2^{k+1}} + \left(x - \sqrt{x^2 - 1}\right)^{2^{k+1}}\right].$$

Using assumption (57.1),

$$
\begin{aligned}
&f^{k+1}(x)\\
&= \left(f^k \circ f\right)(x)\\
&= \frac{1}{2}\left[\left(2x^2 - 1 + \sqrt{(2x^2 - 1)^2 - 1}\right)^{2^k} + \left(2x^2 - 1 - \sqrt{(2x^2 - 1)^2 - 1}\right)^{2^k}\right]\\
&= \frac{1}{2}\left[\left(2x^2 - 1 + 2x\sqrt{x^2 - 1}\right)^{2^k} + \left(2x^2 - 1 - 2x\sqrt{x^2 - 1}\right)^{2^k}\right].
\end{aligned}
$$

Note that

$$2x^2 - 1 + 2x\sqrt{x^2 - 1} = \left(x + \sqrt{x^2 - 1}\right)^2$$

and

$$2x^2 - 1 - 2x\sqrt{x^2 - 1} = \left(x - \sqrt{x^2 - 1}\right)^2.$$

Hence,

$$f^{k+1}(x) = \frac{1}{2}\left[\left(x + \sqrt{x^2 - 1}\right)^{2^{k+1}} + \left(x - \sqrt{x^2 - 1}\right)^{2^{k+1}}\right].$$

Thus, we proved that it is true for $n = k + 1$ if it is true for $n = k$.

Conclusion

We proved that it is true for $n = 1$. Then, we proved that it is true for $n = k +$ if it is true for $n = k$. Therefore, by weak induction,

$$f^n(x) = \frac{1}{2}\left[\left(x + \sqrt{x^2 - 1}\right)^{2^n} + \left(x - \sqrt{x^2 - 1}\right)^{2^n}\right]$$

for all $n \in \mathbb{N}$.

Problem 58

Topic: Iterated Function, Derivative

Problem: If $f(x) = e^x$, then prove

$$\frac{d}{dx} f^n(x) = \prod_{j=1}^{n} f^j(x)$$

for all $n \in \mathbb{N}$ where $f^n(x)$ denotes the n-th iterate of $f(x)$.

Definition - Iterated Function: The n-th iterate of $f(x)$ is the composition of $f(x)$ with itself n times, i.e.

$$f^n(x) = (\underbrace{f \circ f \circ \cdots \circ f}_{n \text{ times}})(x).$$

Solution

Base Cases

We need to prove that it is true for $n = 1$. We can check that

$$\frac{d}{dx} f^1(x) = \frac{d}{dx} f(x) = \frac{d}{dx} e^x = e^x = f^1(x) = \prod_{j=1}^{1} f^j(x),$$

so it is true for $n = 1$.

Induction Step

Assume

$$\frac{d}{dx}f^k(x) = \prod_{j=1}^{k} f^j(x) \tag{58.1}$$

for some $k \geq 1$. To prove the induction step, we need to prove

$$\frac{d}{dx}f^{k+1}(x) = \prod_{j=1}^{k+1} f^j(x).$$

By applying the chain rule of derivative and using assumption (58.1),

$$\frac{d}{dx}f^{k+1}(x) = \frac{d}{dx}\left(f^k \circ f\right)(x)$$

$$= \frac{d}{dx}f^k(e^x)$$

$$= \prod_{j=1}^{k} f^j(e^x) \times \frac{d}{dx}e^x$$

$$= \prod_{j=1}^{k} f^j(e^x) \times e^x.$$

Since $f^j(e^x) = \left(f^j \circ f\right)(x) = f^{j+1}(x)$ and $e^x = f^1(x)$,

$$\frac{d}{dx}f^{k+1}(x) = \prod_{j=1}^{k} f^{j+1}(x) \times f^1(x)$$

$$= \prod_{j=2}^{k+1} f^j(x) \times f^1(x)$$

$$= \prod_{j=1}^{k+1} f^j(x).$$

Thus, we proved that it is true for $n = k + 1$ if it is true for $n = k$.

Conclusion

We proved that it is true for $n = 1$. Then, we proved that it is true for $n = k + 1$ if it is true for $n = k$. Therefore, by weak induction,

$$\frac{d}{dx}f^n(x) = \prod_{j=1}^{n} f^j(x)$$

for all $n \in \mathbb{N}$.

Problem 59

Topic: Iterated Function, Derivative

Problem: If $f(x) = \ln(x)$, then prove

$$\frac{d}{dx} f^n(x) = \frac{1}{x} \left(\prod_{j=1}^{n-1} f^j(x) \right)^{-1}$$

for $n \geq 2$ where $f^n(x)$ denotes the n-th iterate of $f(x)$.

Definition - Iterated Function: The n-th iterate of $f(x)$ is the composition of $f(x)$ with itself n times, i.e.

$$f^n(x) = (\underbrace{f \circ f \circ \cdots \circ f}_{n \text{ times}})(x).$$

Solution

Base Cases

We need to prove that it is true for $n = 2$. Applying the chain rule of derivative,

$$\frac{d}{dx} f^2(x) = \frac{d}{dx} \ln(\ln(x)) = \frac{1}{\ln(x)} \times \frac{1}{x} = \frac{1}{f^1(x)} \times \frac{1}{x} = \frac{1}{x} \left(\prod_{j=1}^{2-1} f^j(x) \right)^{-1},$$

so it is true for $n = 2$.

Induction Step

Assume

$$\frac{d}{dx} f^k(x) = \frac{1}{x} \left(\prod_{j=1}^{k-1} f^j(x) \right)^{-1}$$

(59.1)

for some $k \geq 2$. To prove the induction step, we need to prove

$$\frac{d}{dx} f^{k+1}(x) = \frac{1}{x} \left(\prod_{j=1}^{k} f^j(x) \right)^{-1}.$$

By applying the chain rule of derivative and using assumption (59.1),

$$\frac{d}{dx} f^{k+1}(x) = \frac{d}{dx} \left(f^k \circ f \right)(x)$$

$$= \frac{d}{dx} f^k(\ln(x))$$

$$= \frac{1}{\ln(x)} \left(\prod_{j=1}^{k-1} f^j(\ln(x)) \right)^{-1} \times \frac{d}{dx} \ln(x)$$

$$= \frac{1}{\ln(x)} \left(\prod_{j=1}^{k-1} f^j(\ln(x)) \right)^{-1} \times \frac{1}{x}.$$

Since $f^j(\ln(x)) = \left(f^j \circ f \right)(x) = f^{j+1}(x)$ and $\ln(x) = f^1(x)$,

$$\frac{d}{dx} f^{k+1}(x) = \frac{1}{f^1(x)} \left(\prod_{j=1}^{k-1} f^{j+1}(x) \right)^{-1} \times \frac{1}{x}$$

$$= \frac{1}{f^1(x)} \left(\prod_{j=2}^{k} f^j(x) \right)^{-1} \times \frac{1}{x}$$

$$= \frac{1}{x} \left(\prod_{j=1}^{k} f^j(x) \right)^{-1}.$$

Thus, we proved that it is true for $n = k + 1$ if it is true for $n = k$.

Conclusion

We proved that it is true for $n = 2$. Then, we proved that it is true for $n = k + 1$ if it is true for $n = k$. Therefore, by weak induction,

$$\frac{d}{dx} f^n(x) = \frac{1}{x} \left(\prod_{j=1}^{n-1} f^j(x) \right)^{-1}$$

for $n \geq 2$.

Problem 60

Topic: Iterated Function

Problem: If
$$f(x) = \frac{x^2}{2x - 1},$$
then prove
$$f^n(x) = x^{2^n} \left(\prod_{j=0}^{n-1} \left[x^{2^j} + (x-1)^{2^j} \right] \right)^{-1}$$
for all $n \in \mathbb{N}$ where $f^n(x)$ denotes the n-th iterate of $f(x)$.

Definition - Iterated Function: The n-th iterate of $f(x)$ is the composition of $f(x)$ with itself n times, i.e.
$$f^n(x) = \underbrace{(f \circ f \circ \cdots \circ f)}_{n \text{ times}}(x).$$

Solution

Base Cases

We need to prove that it is true for $n = 1$. We can check that
$$f^1(x) = x^{2^1} \left(\prod_{j=0}^{1-1} \left[x^{2^j} + (x-1)^{2^j} \right] \right)^{-1}$$
$$= x^2 \left[x^{2^0} + (x-1)^{2^0} \right]^{-1} = \frac{x^2}{2x - 1} = f(x),$$

so it is true for $n = 1$.

Induction Step

Assume

$$f^k(x) = x^{2^k} \left(\prod_{j=0}^{k-1} \left[x^{2^j} + (x-1)^{2^j} \right] \right)^{-1} \tag{60.1}$$

for some $k \geq 1$. To prove the induction step, we need to prove

$$f^{k+1}(x) = x^{2^{k+1}} \left(\prod_{j=0}^{k} \left[x^{2^j} + (x-1)^{2^j} \right] \right)^{-1}.$$

Using assumption (60.1),

$$\begin{aligned}
&f^{k+1}(x) \\
&= \left(f^k \circ f \right)(x) \\
&= \left(\frac{x^2}{2x-1} \right)^{2^k} \left(\prod_{j=0}^{k-1} \left[\left(\frac{x^2}{2x-1} \right)^{2^j} + \left(\frac{x^2}{2x-1} - 1 \right)^{2^j} \right] \right)^{-1} \\
&= \left(\frac{2x-1}{2x-1} \right)^{2^k-1} \left(\frac{x^2}{2x-1} \right)^{2^k} \left(\prod_{j=0}^{k-1} \left[\left(\frac{x^2}{2x-1} \right)^{2^j} + \left(\frac{x^2}{2x-1} - 1 \right)^{2^j} \right] \right)^{-1} \\
&= \frac{1}{(2x-1)^{2^k-1}} \cdot \frac{x^{2^{k+1}}}{2x-1} \left(\prod_{j=0}^{k-1} \left[\left(\frac{x^2}{2x-1} \right)^{2^j} + \left(\frac{x^2}{2x-1} - 1 \right)^{2^j} \right] \right)^{-1}.
\end{aligned}$$

In problem 11, we proved that

$$\sum_{j=0}^{n-1} a^j = \frac{a^n - 1}{a - 1}.$$

Let $a = 2$ and $n = k$,

$$2^k - 1 = \sum_{j=0}^{k-1} 2^j = 2^0 + 2^1 + \cdots + 2^{k-2} + 2^{k-1}.$$

So,

$$\begin{aligned}
\frac{1}{(2x-1)^{2^k-1}} &= \frac{1}{(2x-1)^{2^0}} \cdot \frac{1}{(2x-1)^{2^1}} \cdots \frac{1}{(2x-1)^{2^{k-2}}} \cdot \frac{1}{(2x-1)^{2^{k-1}}} \\
&= \left(\prod_{j=0}^{k-1} (2x-1)^{2^j} \right)^{-1}.
\end{aligned}$$

Hence,

$$f^{k+1}(x)$$

$$= \frac{x^{2^{k+1}}}{2x-1} \left(\prod_{j=0}^{k-1} \left[\left(\frac{x^2}{2x-1} \right)^{2^j} + \left(\frac{x^2}{2x-1} - 1 \right)^{2^j} \right] \right)^{-1} \left(\prod_{j=0}^{k-1} (2x-1)^{2^j} \right)^{-1}$$

$$= \frac{x^{2^{k+1}}}{2x-1} \left(\prod_{j=0}^{k-1} \left[(x^2)^{2^j} + (x^2 - 2x + 1)^{2^j} \right] \right)^{-1}$$

$$= \frac{x^{2^{k+1}}}{2x-1} \left(\prod_{j=0}^{k-1} \left[x^{2^{j+1}} + (x-1)^{2^{j+1}} \right] \right)^{-1}$$

$$= \frac{x^{2^{k+1}}}{2x-1} \left(\prod_{j=1}^{k} \left[x^{2^j} + (x-1)^{2^j} \right] \right)^{-1}.$$

Since

$$\frac{1}{2x-1} = \left[x^{2^0} + (x-1)^{2^0} \right]^{-1},$$

we have

$$f^{k+1}(x) = x^{2^{k+1}} \left(\prod_{j=0}^{k} \left[x^{2^j} + (x-1)^{2^j} \right] \right)^{-1}.$$

Thus, we proved that it is true for $n = k + 1$ if it is true for $n = k$.

Conclusion

We proved that it is true for $n = 1$. Then, we proved that it is true for $n = k + 1$ if it is true for $n = k$. Therefore, by weak induction,

$$f^n(x) = x^{2^n} \left(\prod_{j=0}^{n-1} \left[x^{2^j} + (x-1)^{2^j} \right] \right)^{-1}$$

for all $n \in \mathbb{N}$.

Problem 61

Topic: Product

Problem: Prove
$$\prod_{j=1}^{n} \left(1 + \frac{1}{j}\right)^{j} = \frac{(n+1)^{n}}{n!}.$$

Definition - Factorial:
$$n! = n \cdot (n-1) \cdot (n-2) \cdots 2 \cdot 1.$$
$$0! = 1.$$

Solution

Base Cases

We need to prove that it is true for $n = 1$. We can check that
$$\prod_{j=1}^{1} \left(1 + \frac{1}{j}\right)^{j} = \left(1 + \frac{1}{1}\right)^{1} = 2 = \frac{(1+1)^{1}}{1!},$$

so it is true for $n = 1$.

Induction Step

Assume
$$\prod_{j=1}^{k} \left(1 + \frac{1}{j}\right)^{j} = \frac{(k+1)^{k}}{k!} \tag{61.1}$$

for some $k \geq 1$. To prove the induction step, we need to prove

$$\prod_{j=1}^{k+1} \left(1 + \frac{1}{j}\right)^j = \frac{(k+2)^{k+1}}{(k+1)!}.$$

Using assumption (61.1),

$$\begin{aligned}
\prod_{j=1}^{k+1} \left(1 + \frac{1}{j}\right)^j &= \left(1 + \frac{1}{k+1}\right)^{k+1} \times \prod_{j=1}^{k} \left(1 + \frac{1}{j}\right)^j \\
&= \left(\frac{k+2}{k+1}\right)^{k+1} \times \frac{(k+1)^k}{k!} \\
&= \frac{(k+2)^{k+1}}{(k+1) \cdot k!} \\
&= \frac{(k+2)^{k+1}}{(k+1)!}.
\end{aligned}$$

Thus, we proved that it is true for $n = k+1$ if it is true for $n = k$.

Conclusion

We proved that it is true for $n = 1$. Then, we proved that it is true for $n = k+1$ if it is true for $n = k$. Therefore, by weak induction,

$$\prod_{j=1}^{n} \left(1 + \frac{1}{j}\right)^j = \frac{(n+1)^n}{n!}$$

for all $n \in \mathbb{N}$.

Problem 62

Topic: Product, Trigonometry

Problem: Prove
$$\prod_{j=1}^{n} \cos\left(\frac{x}{2^j}\right) = \frac{\sin x}{2^n \sin\left(\frac{x}{2^n}\right)}.$$

Solution

Base Cases

We need to prove that it is true for $n = 1$. Using the identity $\sin(2\theta) = 2\sin(\theta)\cos(\theta)$,

$$\sin(x) = 2\sin\left(\frac{x}{2}\right)\cos\left(\frac{x}{2}\right),$$

so

$$\prod_{j=1}^{1} \cos\left(\frac{x}{2^j}\right) = \cos\left(\frac{x}{2}\right) = \frac{\sin x}{2\sin\left(\frac{x}{2}\right)}.$$

So, it is true for $n = 1$.

Induction Step

Assume

$$\prod_{j=1}^{k} \cos\left(\frac{x}{2^j}\right) = \frac{\sin x}{2^k \sin\left(\frac{x}{2^k}\right)} \tag{62.1}$$

for some $k \geq 1$. To prove the induction step, we need to prove

$$\prod_{j=1}^{k+1} \cos\left(\frac{x}{2^j}\right) = \frac{\sin x}{2^{k+1} \sin\left(\frac{x}{2^{k+1}}\right)}.$$

Using assumption (62.1),

$$\prod_{j=1}^{k+1} \cos\left(\frac{x}{2^j}\right) = \cos\left(\frac{x}{2^{k+1}}\right) \times \prod_{j=1}^{k} \cos\left(\frac{x}{2^j}\right)$$

$$= \cos\left(\frac{x}{2^{k+1}}\right) \times \frac{\sin x}{2^k \sin\left(\frac{x}{2^k}\right)}.$$

Using the identity $\sin(2\theta) = 2\sin(\theta)\cos(\theta)$,

$$\sin\left(\frac{x}{2^k}\right) = 2\sin\left(\frac{x}{2^{k+1}}\right)\cos\left(\frac{x}{2^{k+1}}\right) \Leftrightarrow \cos\left(\frac{x}{2^{k+1}}\right) = \frac{\sin\left(\frac{x}{2^k}\right)}{2\sin\left(\frac{x}{2^{k+1}}\right)},$$

so

$$\prod_{j=1}^{k+1} \cos\left(\frac{x}{2^j}\right) = \frac{\sin\left(\frac{x}{2^k}\right)}{2\sin\left(\frac{x}{2^{k+1}}\right)} \times \frac{\sin x}{2^k \sin\left(\frac{x}{2^k}\right)}$$

$$= \frac{\sin x}{2^{k+1} \sin\left(\frac{x}{2^{k+1}}\right)}.$$

Thus, we proved that it is true for $n = k+1$ if it is true for $n = k$.

Conclusion

We proved that it is true for $n = 1$. Then, we proved that it is true for $n = k+1$ if it is true for $n = k$. Therefore, by weak induction,

$$\prod_{j=1}^{n} \cos\left(\frac{x}{2^j}\right) = \frac{\sin x}{2^n \sin\left(\frac{x}{2^n}\right)}$$

for all $n \in \mathbb{N}$.

Problem 63

Topic: Product, Series

Problem: Prove

$$\prod_{j=0}^{n-1} \left(1 + x^{2^j}\right) = \sum_{j=0}^{2^n-1} x^j.$$

Solution

Base Cases

We need to prove that it is true for $n = 1$. We can check that

$$\prod_{j=0}^{1-1} \left(1 + x^{2^j}\right) = 1 + x^{2^0} = 1 + x = x^0 + x^1 = \sum_{j=0}^{2^1-1} x^j,$$

so it is true for $n = 1$.

Induction Step

Assume

$$\prod_{j=0}^{k-1} \left(1 + x^{2^j}\right) = \sum_{j=0}^{2^k-1} x^j \tag{63.1}$$

for some $k \geq 1$. To prove the induction step, we need to prove

$$\prod_{j=0}^{k} \left(1 + x^{2^j}\right) = \sum_{j=0}^{2^{k+1}-1} x^j.$$

Using assumption (63.1),

$$\prod_{j=0}^{k}\left(1+x^{2^j}\right) = \prod_{j=0}^{k-1}\left(1+x^{2^j}\right) \times \left(1+x^{2^k}\right)$$

$$= \left(\sum_{j=0}^{2^k-1} x^j\right) \times \left(1+x^{2^k}\right)$$

$$= \sum_{j=0}^{2^k-1} x^j + \sum_{j=0}^{2^k-1} x^{j+2^k}.$$

Changing the index $j + 2^k \to j$,

$$\sum_{j=0}^{2^k-1} x^{j+2^k} = \sum_{j=2^k}^{2^k-1+2^k} x^j = \sum_{j=2^k}^{2^{k+1}-1} x^j.$$

Hence,

$$\prod_{j=0}^{k}\left(1+x^{2^j}\right) = \sum_{j=0}^{2^k-1} x^j + \sum_{j=2^k}^{2^{k+1}-1} x^j = \sum_{j=0}^{2^{k+1}-1} x^j.$$

Thus, we proved that it is true for $n = k + 1$ if it is true for $n = k$.

Conclusion

We proved that it is true for $n = 1$. Then, we proved that it is true for $n = k + 1$ if it is true for $n = k$. Therefore, by weak induction,

$$\prod_{j=0}^{n-1}\left(1+x^{2^j}\right) = \sum_{j=0}^{2^n-1} x^j$$

for all $n \in \mathbb{N}$.

Problem 64

Topic: Product, Series

Problem: Prove
$$\prod_{j=0}^{n-1} \left(1 + x^{3^j} + x^{2\cdot 3^j}\right) = \sum_{j=0}^{3^n - 1} x^j.$$

Solution

Base Cases

We need to prove that it is true for $n = 1$. We can check that

$$\prod_{j=0}^{1-1} \left(1 + x^{3^j} + x^{2\cdot 3^j}\right) = 1 + x^{3^0} + x^{2\cdot 3^0} = 1 + x + x^2 = x^0 + x^1 + x^2 = \sum_{j=0}^{3^1 - 1} x^j,$$

so it is true for $n = 1$.

Induction Step

Assume

$$\prod_{j=0}^{k-1} \left(1 + x^{3^j} + x^{2\cdot 3^j}\right) = \sum_{j=0}^{3^k - 1} x^j \tag{64.1}$$

for some $k \geq 1$. To prove the induction step, we need to prove

$$\prod_{j=0}^{k} \left(1 + x^{3^j} + x^{2\cdot 3^j}\right) = \sum_{j=0}^{3^{k+1} - 1} x^j.$$

Using assumption (64.1),

$$\prod_{j=0}^{k}\left(1+x^{3^j}+x^{2\cdot3^j}\right)=\prod_{j=0}^{k-1}\left(1+x^{3^j}+x^{2\cdot3^j}\right)\times\left(1+x^{3^k}+x^{2\cdot3^k}\right)$$

$$=\left(\sum_{j=0}^{3^k-1}x^j\right)\times\left(1+x^{3^k}+x^{2\cdot3^k}\right)$$

$$=\sum_{j=0}^{3^k-1}x^j+\sum_{j=0}^{3^k-1}x^{j+3^k}+\sum_{j=0}^{3^k-1}x^{j+2\cdot3^k}.$$

Changing the index $j+3^k\to j$,

$$\sum_{j=0}^{3^k-1}x^{j+3^k}=\sum_{j=3^k}^{3^k-1+3^k}x^j=\sum_{j=3^k}^{2\cdot3^k-1}x^j.$$

Changing the index $j+2\cdot3^k\to j$,

$$\sum_{j=0}^{3^k-1}x^{j+2\cdot3^k}=\sum_{j=2\cdot3^k}^{3^k-1+2\cdot3^k}x^j=\sum_{j=2\cdot3^k}^{3^{k+1}-1}x^j.$$

Hence,

$$\prod_{j=0}^{k}\left(1+x^{3^j}+x^{2\cdot3^j}\right)=\sum_{j=0}^{3^k-1}x^j+\sum_{j=3^k}^{2\cdot3^k-1}x^j+\sum_{j=2\cdot3^k}^{3^{k+1}-1}x^j=\sum_{j=0}^{3^{k+1}-1}x^j.$$

Thus, we proved that it is true for $n=k+1$ if it is true for $n=k$.

Conclusion

We proved that it is true for $n=1$. Then, we proved that it is true for $n=k+1$ if it is true for $n=k$. Therefore, by weak induction,

$$\prod_{j=0}^{n-1}\left(1+x^{3^j}+x^{2\cdot3^j}\right)=\sum_{j=0}^{3^n-1}x^j$$

for all $n\in\mathbb{N}$.

Problem 65

Topic: Supergolden Number, Narayana Sequence

Problem: Prove that

$$\psi^n = N_{n-2}\psi^2 + N_{n-4}\psi + N_{n-3}$$

for $n \geq 4$ where ψ is the supergolden number and N_n denotes the n-th number in Narayana sequence.

Definition - Supergolden Number: The supergolden number ψ is the real solution of $x^3 - x^2 - 1 = 0$. Its exact value is

$$\psi = \frac{1}{3}\left(1 + \sqrt[3]{\frac{29 + 3\sqrt{93}}{2}} + \sqrt[3]{\frac{29 - 3\sqrt{93}}{2}}\right).$$

Definition - Narayana Sequence: The Narayana sequence $\{N_n\}$ is a sequence of numbers defined by $N_0 = N_1 = N_2 = 1$ and $N_n = N_{n-1} + N_{n-3}$ for $n \geq 3$.

Solution

Base Cases

We need to prove that it is true for $n = 4$. From the definition of supergolden number,

$$\psi^3 - \psi^2 - 1 = 0 \Leftrightarrow \psi^3 = \psi^2 + 1.$$

148

Multiplying both sides by ψ,

$$\psi^4 = \psi^3 + \psi = \psi^2 + \psi + 1 = N_2\psi^2 + N_0\psi + N_1$$

because $N_0 = N_1 = N_2 = 1$. So, it is true for $n = 4$.

Induction Step

Assume

$$\psi^k = N_{k-2}\psi^2 + N_{k-4}\psi + N_{k-3} \tag{65.1}$$

for some $k \geq 4$. To prove the induction step, we need to prove

$$\psi^{k+1} = N_{k-1}\psi^2 + N_{k-3}\psi + N_{k-2}.$$

Multiplying both sides of (65.1) by ψ,

$$\psi^{k+1} = N_{k-2}\psi^3 + N_{k-4}\psi^2 + N_{k-3}\psi.$$

From the definition of supergolden number,

$$\psi^3 - \psi^2 - 1 = 0 \Leftrightarrow \psi^3 = \psi^2 + 1.$$

So,

$$\begin{aligned}
\psi^{k+1} &= N_{k-2}\left(\psi^2 + 1\right) + N_{k-4}\psi^2 + N_{k-3}\psi \\
&= \left(N_{k-2} + N_{k-4}\right)\psi^2 + N_{k-3}\psi + N_{k-2}.
\end{aligned}$$

Using the recursive relation of N_n,

$$N_{k-2} + N_{k-4} = N_{k-1}.$$

Hence,

$$\psi^{k+1} = N_{k-1}\psi^2 + N_{k-3}\psi + N_{k-2}.$$

Thus, we proved that it is true for $n = k + 1$ if it is true for $n = k$.

Conclusion

We proved that it is true for $n = 4$. Then, we proved that it is true for $n = k +$ if it is true for $n = k$. Therefore, by weak induction,

$$\psi^n = N_{n-2}\psi^2 + N_{n-4}\psi + N_{n-3}$$

for $n \geq 4$.

Problem 66

Topic: Inequality

Problem: Prove

$$\left| \sum_{j=1}^{n} x_j \right| \leq \sum_{j=1}^{n} |x_j|$$

for all $n \in \mathbb{N}$ for real numbers x_j, which is an extension of the triangle inequality.

Solution

Base Cases

We need to prove that it is true for $n = 1$ and $n = 2$. When $n = 1$,

$$\left| \sum_{j=1}^{1} x_j \right| = |x_1| \leq |x_1| = \sum_{j=1}^{1} |x_j|,$$

so it is true for $n = 1$. Now using the fact that a number is less than or equal to its absolute value,

$$x_1 x_2 \leq |x_1 x_2|$$
$$x_1 x_2 \leq |x_1||x_2|.$$

Multiplying both sides by 2,

$$2x_1 x_2 \leq 2|x_1||x_2|.$$

Adding $x_1^2 + x_2^2$ to both sides,

$$x_1^2 + 2x_1 x_2 + x_2^2 \leq x_1^2 + 2|x_1||x_2| + x_2^2.$$

Since $x_1^2 = |x_1|^2$ and $x_2^2 = |x_2|^2$,

$$x_1^2 + 2x_1x_2 + x_2^2 \le |x_1|^2 + 2|x_1||x_2| + |x_2|^2$$
$$(x_1 + x_2)^2 \le (|x_1| + |x_2|)^2 .$$

Taking square root of both sides,

$$|x_1 + x_2| \le |x_1| + |x_2|.$$

So, it is true for $n = 2$.

Induction Step

Assume

$$\left| \sum_{j=1}^{m} x_j \right| \le \sum_{j=1}^{m} |x_j| \tag{66.1}$$

for all $1 \le m \le k$ for some $k \ge 2$. To prove the induction step, we need to prove

$$\left| \sum_{j=1}^{k+1} x_j \right| \le \sum_{j=1}^{k+1} |x_j|.$$

By assumption (66.1) when $m = 2$,

$$|x_1 + x_2| \le |x_1| + |x_2|.$$

Let $x_1 = \sum_{j=1}^{k} x_j$ and $x_2 = x_{k+1}$,

$$\left| \sum_{j=1}^{k} x_j + x_{k+1} \right| \le \left| \sum_{j=1}^{k} x_j \right| + |x_{k+1}|$$

$$\left| \sum_{j=1}^{k+1} x_j \right| \le \left| \sum_{j=1}^{k} x_j \right| + |x_{k+1}|.$$

By assumption (66.1) when $m = k$,

$$\left| \sum_{j=1}^{k} x_j \right| \le \sum_{j=1}^{k} |x_j|.$$

Hence,

$$\left| \sum_{j=1}^{k+1} x_j \right| \le \sum_{j=1}^{k} |x_j| + |x_{k+1}| = \sum_{j=1}^{k+1} |x_j|.$$

Thus, we proved that it is true for $n = k+1$ if it is true for $n = m$ for all $1 \le m \le k$

Conclusion

We proved that it is true for $n = 1$ and $n = 2$. Then, we proved that it is true for $n = k + 1$ if it is true for $n = m$ for all $1 \leq m \leq k$. Therefore, by strong induction,

$$\left| \sum_{j=1}^{n} x_j \right| \leq \sum_{j=1}^{n} |x_j|$$

for all $n \in \mathbb{N}$.

Problem 67

Topic: Inequality

Problem: Prove Cauchy-Schwarz inequality,

$$(a_1 b_1 + a_2 b_2 + \cdots + a_n b_n)^2 \leq \left(a_1^2 + a_2^2 + \cdots + a_n^2\right)\left(b_1^2 + b_2^2 + \cdots + b_n^2\right),$$

for all $n \in \mathbb{N}$ and $a_1, \cdots, a_n, b_1, \cdots, b_n \geq 0$.

Solution

Base Cases

We need to prove that it is true for $n = 1$ and $n = 2$. When $n = 1$,

$$(a_1 b_1)^2 = a_1^2 b_1^2 \leq \left(a_1^2\right)\left(b_1^2\right),$$

so it is true for $n = 1$. When $n = 2$,

$$\left(a_1^2 + a_2^2\right)\left(b_1^2 + b_2^2\right) - (a_1 b_1 + a_2 b_2)^2 = a_1^2 b_2^2 + a_2^2 b_1^2 - 2 a_1 b_1 a_2 b_2$$
$$= (a_1 b_2 - a_2 b_1)^2 \geq 0,$$

so it is true that

$$(a_1 b_1 + a_2 b_2)^2 \leq \left(a_1^2 + a_2^2\right)\left(b_1^2 + b_2^2\right).$$

Induction Step

Assume

$$(a_1 b_1 + a_2 b_2 + \cdots + a_m b_m)^2 \leq \left(a_1^2 + a_2^2 + \cdots + a_m^2\right)\left(b_1^2 + b_2^2 + \cdots + b_m^2\right)$$
$$\Leftrightarrow a_1 b_1 + a_2 b_2 + \cdots + a_m b_m \leq \sqrt{a_1^2 + a_2^2 + \cdots + a_m^2}\sqrt{b_1^2 + b_2^2 + \cdots + b_m^2} \quad (67.1$$

for all $1 \leq m \leq k$ for some $k \geq 2$. To prove the induction step, we need to prove

$$(a_1 b_1 + a_2 b_2 + \cdots + a_{k+1} b_{k+1})^2 \leq \left(a_1^2 + a_2^2 + \cdots + a_{k+1}^2\right)\left(b_1^2 + b_2^2 + \cdots + b_{k+1}^2\right).$$

By assumption (67.1) when $m = k$,

$$a_1 b_1 + a_2 b_2 + \cdots + a_k b_k + a_{k+1} b_{k+1}$$
$$\leq \sqrt{a_1^2 + a_2^2 + \cdots + a_k^2}\sqrt{b_1^2 + b_2^2 + \cdots + b_k^2} + a_{k+1} b_{k+1}. \tag{67.2}$$

By assumption (67.1) when $m = 2$,

$$a_1 b_1 + a_2 b_2 \leq \sqrt{a_1^2 + a_2^2}\sqrt{b_1^2 + b_2^2}.$$

Let $a_1 = \sqrt{a_1^2 + a_2^2 + \cdots + a_k^2}$, $b_1 = \sqrt{b_1^2 + b_2^2 + \cdots + b_k^2}$, $a_2 = a_{k+1}$, and $b_2 = b_{k+1}$,

$$\sqrt{a_1^2 + a_2^2 + \cdots + a_k^2}\sqrt{b_1^2 + b_2^2 + \cdots + b_k^2} + a_{k+1} b_{k+1}$$
$$\leq \sqrt{a_1^2 + a_2^2 + \cdots + a_k^2 + a_{k+1}^2}\sqrt{b_1^2 + b_2^2 + \cdots + b_k^2 + b_{k+1}^2}. \tag{67.3}$$

Hence, from (67.2) and (67.3),

$$a_1 b_1 + a_2 b_2 + \cdots + a_{k+1} b_{k+1} \leq \sqrt{a_1^2 + a_2^2 + \cdots + a_{k+1}^2}\sqrt{b_1^2 + b_2^2 + \cdots + b_{k+1}^2}$$

$$(a_1 b_1 + a_2 b_2 + \cdots + a_{k+1} b_{k+1})^2 \leq \left(a_1^2 + a_2^2 + \cdots + a_{k+1}^2\right)\left(b_1^2 + b_2^2 + \cdots + b_{k+1}^2\right).$$

Thus, we proved that it is true for $n = k+1$ if it is true for $n = m$ for all $1 \leq m \leq k$.

Conclusion

We proved that it is true for $n = 1$ and $n = 2$. Then, we proved that it is true for $n = k + 1$ if it is true for $n = m$ for all $1 \leq m \leq k$. Therefore, by strong induction,

$$(a_1 b_1 + a_2 b_2 + \cdots + a_n b_n)^2 \leq \left(a_1^2 + a_2^2 + \cdots + a_n^2\right)\left(b_1^2 + b_2^2 + \cdots + b_n^2\right)$$

for all $n \in \mathbb{N}$.

Problem 68

Topic: Inequality

Problem: Prove Aczél's inequality, which states that if

$$x_1^2 + x_2^2 + \cdots + x_n^2 < x_0^2$$

and

$$y_1^2 + y_2^2 + \cdots + y_n^2 < y_0^2,$$

then

$$\left(x_0^2 - x_1^2 - x_2^2 - \cdots - x_n^2\right)\left(y_0^2 - y_1^2 - y_2^2 - \cdots - y_n^2\right)$$
$$\leq \left(x_0 y_0 - x_1 y_1 - x_2 y_2 - \cdots - x_n y_n\right)^2.$$

Solution

Base Cases

We need to prove that it is true for $n = 1$, i.e. if $x_1^2 < x_0^2$ and $y_1^2 < y_0^2$, then

$$\left(x_0^2 - x_1^2\right)\left(y_0^2 - y_1^2\right) \leq \left(x_0 y_0 - x_1 y_1\right)^2.$$

Since

$$\left(x_0 y_0 - x_1 y_1\right)^2 - \left(x_0^2 - x_1^2\right)\left(y_0^2 - y_1^2\right) = -2x_0 y_0 x_1 y_1 + x_1^2 y_0^2 + x_0^2 y_1^2$$
$$= \left(x_1 y_0 - x_0 y_1\right)^2 \geq 0,$$

it is true that

$$\left(x_0^2 - x_1^2\right)\left(y_0^2 - y_1^2\right) \leq \left(x_0 y_0 - x_1 y_1\right)^2.$$

Induction Step

Assume if $x_1^2 + x_2^2 + \cdots + x_m^2 < x_0^2$ and $y_1^2 + y_2^2 + \cdots + y_m^2 < y_0^2$, then

$$\left(x_0^2 - x_1^2 - x_2^2 - \cdots - x_m^2\right)\left(y_0^2 - y_1^2 - y_2^2 - \cdots - y_m^2\right)$$
$$\leq \left(x_0 y_0 - x_1 y_1 - x_2 y_2 - \cdots - x_m y_m\right)^2 \tag{68.1}$$

for all $1 \leq m \leq k$ for some $k \geq 1$. To prove the induction step, we need to prove that if $x_1^2 + x_2^2 + \cdots + x_{k+1}^2 < x_0^2$ and $y_1^2 + y_2^2 + \cdots + y_{k+1}^2 < y_0^2$, then

$$\left(x_0^2 - x_1^2 - x_2^2 - \cdots - x_{k+1}^2\right)\left(y_0^2 - y_1^2 - y_2^2 - \cdots - y_{k+1}^2\right)$$
$$\leq \left(x_0 y_0 - x_1 y_1 - x_2 y_2 - \cdots - x_{k+1} y_{k+1}\right)^2 .$$

First, note that

$$x_1^2 + x_2^2 + \cdots + x_{k+1}^2 < x_0^2 \Rightarrow x_1^2 + x_2^2 + \cdots + x_k^2 < \left(\sqrt{x_0^2 - x_{k+1}^2}\right)^2 \tag{68.2}$$

and

$$y_1^2 + y_2^2 + \cdots + y_{k+1}^2 < y_0^2 \Rightarrow y_1^2 + y_2^2 + \cdots + y_k^2 < \left(\sqrt{y_0^2 - y_{k+1}^2}\right)^2 . \tag{68.3}$$

By assumption (68.1) when $m = k$, if $x_1^2 + x_2^2 + \cdots + x_k^2 < x_0^2$ and $y_1^2 + y_2^2 + \cdots + y_k^2 < y_0^2$, then

$$\left(x_0^2 - x_1^2 - x_2^2 - \cdots - x_k^2\right)\left(y_0^2 - y_1^2 - y_2^2 - \cdots - y_k^2\right)$$
$$\leq \left(x_0 y_0 - x_1 y_1 - x_2 y_2 - \cdots - x_k y_k\right)^2 .$$

Let $x_0 = \sqrt{x_0^2 - x_{k+1}^2}$ and $y_0 = \sqrt{y_0^2 - y_{k+1}^2}$, if

$$x_1^2 + x_2^2 + \cdots + x_k^2 < \left(\sqrt{x_0^2 - x_{k+1}^2}\right)^2$$

and

$$y_1^2 + y_2^2 + \cdots + y_k^2 < \left(\sqrt{y_0^2 - y_{k+1}^2}\right)^2 ,$$

then

$$\left(x_0^2 - x_{k+1}^2 - x_1^2 - x_2^2 - \cdots - x_k^2\right)\left(y_0^2 - y_{k+1}^2 - y_1^2 - y_2^2 - \cdots - y_k^2\right)$$
$$\leq \left(\sqrt{x_0^2 - x_{k+1}^2}\sqrt{y_0^2 - y_{k+1}^2} - x_1 y_1 - x_2 y_2 - \cdots - x_k y_k\right)^2 . \tag{68.4}$$

By assumption (68.1) when $m = 1$, if $x_1^2 < x_0^2$ and $y_1^2 < y_0^2$, then

$$\left(x_0^2 - x_1^2\right)\left(y_0^2 - y_1^2\right) \leq \left(x_0 y_0 - x_1 y_1\right)^2 .$$

Let $x_1 = x_{k+1}$ and $y_1 = y_{k+1}$, if $x_{k+1}^2 < x_0^2$ and $y_{k+1}^2 < y_0^2$, then

$$\left(x_0^2 - x_{k+1}^2\right)\left(y_0^2 - y_{k+1}^2\right) \leq \left(x_0 y_0 - x_{k+1} y_{k+1}\right)^2 ,$$

which implies

$$\sqrt{x_0^2 - x_{k+1}^2}\sqrt{y_0^2 - y_{k+1}^2} \leq x_0 y_0 - x_{k+1}y_{k+1}.$$

So,

$$\left(\sqrt{x_0^2 - x_{k+1}^2}\sqrt{y_0^2 - y_{k+1}^2} - x_1 y_1 - x_2 y_2 - \cdots - x_k y_k\right)^2$$
$$\leq \left(x_0 y_0 - x_{k+1}y_{k+1} - x_1 y_1 - x_2 y_2 - \cdots - x_k y_k\right)^2. \tag{68.5}$$

Hence, from (68.2), (68.3), (68.4), and (68.5), if $x_1^2 + x_2^2 + \cdots + x_{k+1}^2 < x_0^2$ and $y_1^2 + y_2^2 + \cdots + y_{k+1}^2 < y_0^2$, then

$$\left(x_0^2 - x_1^2 - x_2^2 - \cdots - x_{k+1}^2\right)\left(y_0^2 - y_1^2 - y_2^2 - \cdots - y_{k+1}^2\right)$$
$$\leq \left(x_0 y_0 - x_1 y_1 - x_2 y_2 - \cdots - x_{k+1}y_{k+1}\right)^2.$$

Thus, we proved that it is true for $n = k+1$ if it is true for $n = m$ for all $1 \leq m \leq k$.

Conclusion

We proved that it is true for $n = 1$. Then, we proved that it is true for $n = k + 1$ if it is true for $n = m$ for all $1 \leq m \leq k$. Therefore, by strong induction, if $x_1^2 + x_2^2 + \cdots + x_n^2 < x_0^2$ and $y_1^2 + y_2^2 + \cdots + y_n^2 < y_0^2$, then

$$\left(x_0^2 - x_1^2 - x_2^2 - \cdots - x_n^2\right)\left(y_0^2 - y_1^2 - y_2^2 - \cdots - y_n^2\right)$$
$$\leq \left(x_0 y_0 - x_1 y_1 - x_2 y_2 - \cdots - x_n y_n\right)^2$$

for all $n \in \mathbb{N}$.

Problem 69

Topic: Inequality

Problem: Prove

$$\sum_{j=1}^{n} a_j - \prod_{j=1}^{n} a_j \leq n - 1$$

if $0 \leq a_j \leq 1$ for all $1 \leq j \leq n$ for all $n \in \mathbb{N}$.

Solution

Base Cases

We need to prove that it is true for $n = 1$. We can check that

$$\sum_{j=1}^{1} a_j - \prod_{j=1}^{1} a_j = a_1 - a_1 = 0 \leq 1 - 1,$$

so it is true for $n = 1$.

Induction Step

Assume

$$\sum_{j=1}^{k} a_j - \prod_{j=1}^{k} a_j \leq k - 1 \tag{69.1}$$

if $0 \leq a_j \leq 1$ for all $1 \leq j \leq k$ for some $k \geq 1$. To prove the induction step, we need to prove

$$\sum_{j=1}^{k+1} a_j - \prod_{j=1}^{k+1} a_j \leq (k + 1) - 1 = k$$

if $0 \leq a_j \leq 1$ for all $1 \leq j \leq k+1$. First, if $0 \leq a_j \leq 1$ for all $1 \leq j \leq k+1$, then $a_1 + a_2 + \cdots + a_k \leq k$ and $a_{k+1} \leq 1$, which imply

$$\left(k - \sum_{j=1}^{k} a_j \right) \geq 0$$

and $(1 - a_{k+1}) \geq 0$. Multiplying them together,

$$\left(k - \sum_{j=1}^{k} a_j \right) (1 - a_{k+1}) \geq 0$$

$$k - \sum_{j=1}^{k} a_j - ka_{k+1} + a_{k+1} \sum_{j=1}^{k} a_j \geq 0$$

$$k - \sum_{j=1}^{k} a_j - a_{k+1} - (k-1)a_{k+1} + a_{k+1} \sum_{j=1}^{k} a_j \geq 0$$

$$k - \left(\sum_{j=1}^{k} a_j + a_{k+1} \right) + a_{k+1} \left(\sum_{j=1}^{k} a_j - (k-1) \right) \geq 0$$

$$k - \sum_{j=1}^{k+1} a_j + a_{k+1} \left(\sum_{j=1}^{k} a_j - (k-1) \right) \geq 0.$$

So,

$$\sum_{j=1}^{k+1} a_j \leq k + a_{k+1} \left(\sum_{j=1}^{k} a_j - (k-1) \right).$$

By assumption (69.1),

$$\sum_{j=1}^{k} a_j - (k-1) \leq \prod_{j=1}^{k} a_j,$$

so

$$k + a_{k+1} \left(\sum_{j=1}^{k} a_j - (k-1) \right) \leq k + a_{k+1} \cdot \prod_{j=1}^{k} a_j$$

$$= k + \prod_{j=1}^{k+1} a_j.$$

Hence,

$$\sum_{j=1}^{k+1} a_j \leq k + \prod_{j=1}^{k+1} a_j$$

$$\sum_{j=1}^{k+1} a_j - \prod_{j=1}^{k+1} a_j \leq k.$$

Thus, we proved that it is true for $n = k+1$ if it is true for $n = k$.

Conclusion

We proved that it is true for $n = 1$. Then, we proved that it is true for $n = k + 1$ if it is true for $n = k$. Therefore, by weak induction,

$$\sum_{j=1}^{n} a_j - \prod_{j=1}^{n} a_j \leq n - 1$$

if $0 \leq a_j \leq 1$ for all $1 \leq j \leq n$ for all $n \in \mathbb{N}$.

Problem 70

Topic: Inequality

Problem: For $x_{1,j}, x_{2,j}, \cdots, x_{n,j} \geq 0$ for all $1 \leq j \leq \ell$, prove

$$\left(\sum_{j=1}^{\ell} x_{1,j} x_{2,j} \cdots x_{n,j} \right)^2 \leq \left(\sum_{j=1}^{\ell} x_{1,j}^2 \right) \left(\sum_{j=1}^{\ell} x_{2,j}^2 \right) \cdots \left(\sum_{j=1}^{\ell} x_{n,j}^2 \right)$$

for $n \geq 2$.

Solution

Base Cases

We need to prove that it is true for $n = 2$. In problem 67, we proved that

$$\left(\sum_{j=1}^{\ell} a_j b_j \right)^2 \leq \left(\sum_{j=1}^{\ell} a_j^2 \right) \left(\sum_{j=1}^{\ell} b_j^2 \right).$$

Let $a_j \to x_{1,j}$ and $b_j \to x_{2,j}$,

$$\left(\sum_{j=1}^{\ell} x_{1,j} x_{2,j} \right)^2 \leq \left(\sum_{j=1}^{\ell} x_{1,j}^2 \right) \left(\sum_{j=1}^{\ell} x_{2,j}^2 \right).$$

So, it is true for $n = 2$.

Induction Step

Assume

$$\left(\sum_{j=1}^{\ell} x_{1,j}x_{2,j}\cdots x_{k,j}\right)^2 \le \left(\sum_{j=1}^{\ell} x_{1,j}^2\right)\left(\sum_{j=1}^{\ell} x_{2,j}^2\right)\cdots\left(\sum_{j=1}^{\ell} x_{k,j}^2\right) \qquad (70.1)$$

for some $k \ge 2$. To prove the induction step, we need to prove

$$\left(\sum_{j=1}^{\ell} x_{1,j}x_{2,j}\cdots x_{k+1,j}\right)^2 \le \left(\sum_{j=1}^{\ell} x_{1,j}^2\right)\left(\sum_{j=1}^{\ell} x_{2,j}^2\right)\cdots\left(\sum_{j=1}^{\ell} x_{k+1,j}^2\right).$$

Using assumption (70.1), let $x_{k,j} \to x_{k,j}x_{k+1,j}$,

$$\left(\sum_{j=1}^{\ell} x_{1,j}x_{2,j}\cdots x_{k,j}x_{k+1,j}\right)^2 \le \left(\sum_{j=1}^{\ell} x_{1,j}^2\right)\left(\sum_{j=1}^{\ell} x_{2,j}^2\right)\cdots\left(\sum_{j=1}^{\ell} x_{k,j}^2 x_{k+1,j}^2\right).$$

By expanding the product of two sums, we have

$$\left(\sum_{j=1}^{\ell} x_{k,j}^2\right)\left(\sum_{j=1}^{\ell} x_{k+1,j}^2\right)$$

$$= \sum_{j=1}^{\ell} x_{k,j}^2 x_{k+1,j}^2 + \sum_{1\le i<j\le \ell} x_{k,i}^2 x_{k+1,j}^2 + \sum_{1\le j<i\le \ell} x_{k,i}^2 x_{k+1,j}^2$$

$$\ge \sum_{j=1}^{\ell} x_{k,j}^2 x_{k+1,j}^2.$$

Hence,

$$\left(\sum_{j=1}^{\ell} x_{1,j}x_{2,j}\cdots x_{k,j}x_{k+1,j}\right)^2$$

$$\le \left(\sum_{j=1}^{\ell} x_{1,j}^2\right)\left(\sum_{j=1}^{\ell} x_{2,j}^2\right)\cdots\left(\sum_{j=1}^{\ell} x_{k,j}^2\right)\left(\sum_{j=1}^{\ell} x_{k+1,j}^2\right).$$

Thus, we proved that it is true for $n = k + 1$ if it is true for $n = k$.

Conclusion

We proved that it is true for $n = 2$. Then, we proved that it is true for $n = k + 1$ if it is true for $n = k$. Therefore, by weak induction,

$$\left(\sum_{j=1}^{\ell} x_{1,j}x_{2,j}\cdots x_{n,j}\right)^2 \le \left(\sum_{j=1}^{\ell} x_{1,j}^2\right)\left(\sum_{j=1}^{\ell} x_{2,j}^2\right)\cdots\left(\sum_{j=1}^{\ell} x_{n,j}^2\right)$$

for $n \ge 2$.

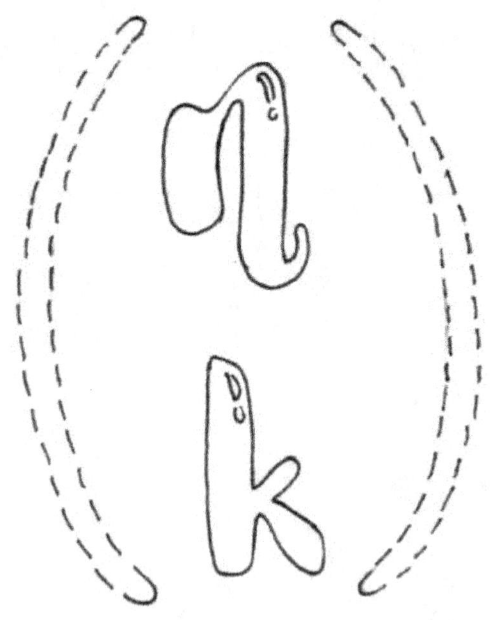

Problem 71

Topic: Inequality

> **Problem:** For $a_j, b_j > 0$ for all $1 \leq j \leq n$, prove Milne's inequality
>
> $$\left(\sum_{j=1}^{n} (a_j + b_j) \right) \left(\sum_{j=1}^{n} \frac{a_j b_j}{a_j + b_j} \right) \leq \left(\sum_{j=1}^{n} a_j \right) \left(\sum_{j=1}^{n} b_j \right)$$
>
> for all $n \in \mathbb{N}$.

Solution

Base Cases

We need to prove that it is true for $n = 1$ and $n = 2$. When $n = 1$, we can check that

$$\left(\sum_{j=1}^{1} (a_j + b_j) \right) \left(\sum_{j=1}^{1} \frac{a_j b_j}{a_j + b_j} \right) = a_1 b_1 \leq a_1 b_1 = \left(\sum_{j=1}^{1} a_j \right) \left(\sum_{j=1}^{1} b_j \right),$$

so it is true for $n = 1$. For $n = 2$, since

$$\left(\sum_{j=1}^{2} a_j \right) \left(\sum_{j=1}^{2} b_j \right) - \left(\sum_{j=1}^{2} (a_j + b_j) \right) \left(\sum_{j=1}^{2} \frac{a_j b_j}{a_j + b_j} \right)$$

$$= a_1 b_2 + a_2 b_1 - \frac{(a_1 + b_1) a_2 b_2}{a_2 + b_2} - \frac{(a_2 + b_2) a_1 b_1}{a_1 + b_1}$$

$$= \frac{(a_1 b_2 - a_2 b_1)^2}{(a_1 + b_1)(a_2 + b_2)} \geq 0,$$

it is true that

$$\left(\sum_{j=1}^{2}(a_j + b_j)\right)\left(\sum_{j=1}^{2}\frac{a_j b_j}{a_j + b_j}\right) \le \left(\sum_{j=1}^{2}a_j\right)\left(\sum_{j=1}^{2}b_j\right).$$

Induction Step

Assume

$$\left(\sum_{j=1}^{m}(a_j + b_j)\right)\left(\sum_{j=1}^{m}\frac{a_j b_j}{a_j + b_j}\right) \le \left(\sum_{j=1}^{m}a_j\right)\left(\sum_{j=1}^{m}b_j\right) \tag{71.1}$$

for all $1 \le m \le k$ for some $k \ge 2$. To prove the induction step, we need to prove

$$\left(\sum_{j=1}^{k+1}(a_j + b_j)\right)\left(\sum_{j=1}^{k+1}\frac{a_j b_j}{a_j + b_j}\right) \le \left(\sum_{j=1}^{k+1}a_j\right)\left(\sum_{j=1}^{k+1}b_j\right).$$

First, note that

$$\left(\sum_{j=1}^{k+1}(a_j + b_j)\right)\left(\sum_{j=1}^{k+1}\frac{a_j b_j}{a_j + b_j}\right)$$

$$= \left(\sum_{j=1}^{k}(a_j + b_j) + (a_{k+1} + b_{k+1})\right)\left(\sum_{j=1}^{k}\frac{a_j b_j}{a_j + b_j} + \frac{a_{k+1}b_{k+1}}{a_{k+1} + b_{k+1}}\right)$$

$$= \left(\sum_{j=1}^{k}(a_j + b_j)\right)\left(\sum_{j=1}^{k}\frac{a_j b_j}{a_j + b_j}\right) + a_{k+1}b_{k+1} + \frac{a_{k+1}b_{k+1}}{a_{k+1} + b_{k+1}}\sum_{j=1}^{k}(a_j + b_j)$$

$$+ (a_{k+1} + b_{k+1})\sum_{j=1}^{k}\frac{a_j b_j}{a_j + b_j}$$

and

$$\left(\sum_{j=1}^{k+1}a_j\right)\left(\sum_{j=1}^{k+1}b_j\right)$$

$$= \left(\sum_{j=1}^{k}a_j + a_{k+1}\right)\left(\sum_{j=1}^{k}b_j + b_{k+1}\right)$$

$$= \left(\sum_{j=1}^{k}a_j\right)\left(\sum_{j=1}^{k}b_j\right) + a_{k+1}b_{k+1} + b_{k+1}\sum_{j=1}^{k}a_j + a_{k+1}\sum_{j=1}^{k}b_j.$$

By assumption (71.1) when $m = k$,

$$\left(\sum_{j=1}^{k}(a_j + b_j)\right)\left(\sum_{j=1}^{k}\frac{a_j b_j}{a_j + b_j}\right) \le \left(\sum_{j=1}^{k}a_j\right)\left(\sum_{j=1}^{k}b_j\right). \tag{71.2}$$

By assumption (71.1) when $m = 2$,

$$[(a_1 + b_1) + (a_2 + b_2)] \left[\frac{a_1 b_1}{a_1 + b_1} + \frac{a_2 b_2}{a_2 + b_2} \right] \le (a_1 + a_2)(b_1 + b_2)$$

$$a_1 b_1 + a_2 b_2 + \frac{(a_1 + b_1)a_2 b_2}{a_2 + b_2} + \frac{(a_2 + b_2)a_1 b_1}{a_1 + b_1} \le a_1 b_1 + a_2 b_2 + a_1 b_2 + a_2 b_1$$

$$\frac{(a_1 + b_1)a_2 b_2}{a_2 + b_2} + \frac{(a_2 + b_2)a_1 b_1}{a_1 + b_1} \le a_1 b_2 + a_2 b_1.$$

Let $a_1 = a_j$, $b_1 = b_j$, $a_2 = a_{k+1}$, $b_2 = b_{k+1}$, and then taking the sum from $j = 1$ to $j = k$,

$$\sum_{j=1}^{k} \left[\frac{(a_j + b_j)a_{k+1}b_{k+1}}{a_{k+1} + b_{k+1}} + \frac{(a_{k+1} + b_{k+1})a_j b_j}{a_j + b_j} \right] \le \sum_{j=1}^{k} (a_j b_{k+1} + a_{k+1} b_j)$$

$$\frac{a_{k+1}b_{k+1}}{a_{k+1} + b_{k+1}} \sum_{j=1}^{k}(a_j + b_j) + (a_{k+1} + b_{k+1}) \sum_{j=1}^{k} \frac{a_j b_j}{a_j + b_j} \le b_{k+1} \sum_{j=1}^{k} a_j + a_{k+1} \sum_{j=1}^{k} b_j.$$

$$\tag{71.3}$$

Adding (71.2) and (71.3) and then adding $a_{k+1}b_{k+1}$ to both sides, we have the inequality

$$\left(\sum_{j=1}^{k+1}(a_j + b_j) \right) \left(\sum_{j=1}^{k+1} \frac{a_j b_j}{a_j + b_j} \right) \le \left(\sum_{j=1}^{k+1} a_j \right) \left(\sum_{j=1}^{k+1} b_j \right).$$

Thus, we proved that it is true for $n = k+1$ if it is true for $n = m$ for all $1 \le m \le k$.

Conclusion

We proved that it is true for $n = 1$ and $n = 2$. Then, we proved that it is true for $n = k+1$ if it is true for $n = m$ for all $1 \le m \le k$. Therefore, by strong induction,

$$\left(\sum_{j=1}^{n}(a_j + b_j) \right) \left(\sum_{j=1}^{n} \frac{a_j b_j}{a_j + b_j} \right) \le \left(\sum_{j=1}^{n} a_j \right) \left(\sum_{j=1}^{n} b_j \right)$$

for all $n \in \mathbb{N}$.

Problem 72

Topic: Inequality

Problem: For $a_j, b_j > 0$ for all $1 \leq j \leq n$, prove Milne's inequality

$$\left(\sum_{j=1}^{n} a_j b_j \right)^2 \leq \left(\sum_{j=1}^{n} (a_j^2 + b_j^2) \right) \left(\sum_{j=1}^{n} \frac{a_j^2 b_j^2}{a_j^2 + b_j^2} \right)$$

for all $n \in \mathbb{N}$.

Solution

Base Cases

We need to prove that it is true for $n = 1$ and $n = 2$. When $n = 1$, we can check that

$$\left(\sum_{j=1}^{1} a_j b_j \right)^2 = a_1^2 b_1^2 \leq a_1^2 b_1^2 = \left(\sum_{j=1}^{1} (a_j^2 + b_j^2) \right) \left(\sum_{j=1}^{1} \frac{a_j^2 b_j^2}{a_j^2 + b_j^2} \right),$$

so it is true for $n = 1$. For $n = 2$, since

$$\left(\sum_{j=1}^{2} (a_j^2 + b_j^2) \right) \left(\sum_{j=1}^{2} \frac{a_j^2 b_j^2}{a_j^2 + b_j^2} \right) - \left(\sum_{j=1}^{2} a_j b_j \right)^2$$

$$= \frac{(a_1^2 + b_1^2)\, a_2^2 b_2^2}{a_2^2 + b_2^2} + \frac{(a_2^2 + b_2^2)\, a_1^2 b_1^2}{a_1^2 + b_1^2} - 2 a_1 b_1 a_2 b_2$$

$$= \frac{\left[(a_1^2 + b_1^2)\, a_2 b_2 - (a_2^2 + b_2^2)\, a_1 b_1 \right]^2}{(a_1^2 + b_1^2)(a_2^2 + b_2^2)} \geq 0,$$

it is true that

$$\left(\sum_{j=1}^{2} a_j b_j\right)^2 \leq \left(\sum_{j=1}^{2} \left(a_j^2 + b_j^2\right)\right)\left(\sum_{j=1}^{2} \frac{a_j^2 b_j^2}{a_j^2 + b_j^2}\right).$$

Induction Step

Assume

$$\left(\sum_{j=1}^{m} a_j b_j\right)^2 \leq \left(\sum_{j=1}^{m} \left(a_j^2 + b_j^2\right)\right)\left(\sum_{j=1}^{m} \frac{a_j^2 b_j^2}{a_j^2 + b_j^2}\right) \tag{72.1}$$

for all $1 \leq m \leq k$ for some $k \geq 2$. To prove the induction step, we need to prove

$$\left(\sum_{j=1}^{k+1} a_j b_j\right)^2 \leq \left(\sum_{j=1}^{k+1} \left(a_j^2 + b_j^2\right)\right)\left(\sum_{j=1}^{k+1} \frac{a_j^2 b_j^2}{a_j^2 + b_j^2}\right).$$

First, note that

$$\left(\sum_{j=1}^{k+1} a_j b_j\right)^2 = \left(\sum_{j=1}^{k} a_j b_j + a_{k+1} b_{k+1}\right)^2$$

$$= \left(\sum_{j=1}^{k} a_j b_j\right)^2 + a_{k+1}^2 b_{k+1}^2 + 2 a_{k+1} b_{k+1} \sum_{j=1}^{k} a_j b_j$$

and

$$\left(\sum_{j=1}^{k+1} \left(a_j^2 + b_j^2\right)\right)\left(\sum_{j=1}^{k+1} \frac{a_j^2 b_j^2}{a_j^2 + b_j^2}\right)$$

$$= \left(\sum_{j=1}^{k} \left(a_j^2 + b_j^2\right) + \left(a_{k+1}^2 + b_{k+1}^2\right)\right)\left(\sum_{j=1}^{k} \frac{a_j^2 b_j^2}{a_j^2 + b_j^2} + \frac{a_{k+1}^2 b_{k+1}^2}{a_{k+1}^2 + b_{k+1}^2}\right)$$

$$= \left(\sum_{j=1}^{k} \left(a_j^2 + b_j^2\right)\right)\left(\sum_{j=1}^{k} \frac{a_j^2 b_j^2}{a_j^2 + b_j^2}\right) + a_{k+1}^2 b_{k+1}^2 + \frac{a_{k+1}^2 b_{k+1}^2}{a_{k+1}^2 + b_{k+1}^2} \sum_{j=1}^{k} \left(a_j^2 + b_j^2\right)$$

$$+ \left(a_{k+1}^2 + b_{k+1}^2\right) \sum_{j=1}^{k} \frac{a_j^2 b_j^2}{a_j^2 + b_j^2}.$$

By assumption (72.1) when $m = k$,

$$\left(\sum_{j=1}^{k} a_j b_j\right)^2 \leq \left(\sum_{j=1}^{k} \left(a_j^2 + b_j^2\right)\right)\left(\sum_{j=1}^{k} \frac{a_j^2 b_j^2}{a_j^2 + b_j^2}\right). \tag{72.2}$$

By assumption (72.1) when $m = 2$,

$$(a_1b_1 + a_2b_2)^2 \leq \left[(a_1^2 + b_1^2) + (a_2^2 + b_2^2)\right] \left[\frac{a_1^2 b_1^2}{a_1^2 + b_1^2} + \frac{a_2^2 b_2^2}{a_2^2 + b_2^2}\right]$$

$$a_1^2 b_1^2 + a_2^2 b_2^2 + 2a_1b_1a_2b_2 \leq a_1^2 b_1^2 + a_2^2 b_2^2 + \frac{\left(a_1^2 + b_1^2\right) a_2^2 b_2^2}{a_2^2 + b_2^2} + \frac{\left(a_2^2 + b_2^2\right) a_1^2 b_1^2}{a_1^2 + b_1^2}$$

$$2a_1b_1a_2b_2 \leq \frac{\left(a_1^2 + b_1^2\right) a_2^2 b_2^2}{a_2^2 + b_2^2} + \frac{\left(a_2^2 + b_2^2\right) a_1^2 b_1^2}{a_1^2 + b_1^2}.$$

Let $a_1 = a_j$, $b_1 = b_j$, $a_2 = a_{k+1}$, $b_2 = b_{k+1}$, and then taking the sum from $j = 1$ to $j = k$,

$$\sum_{j=1}^{k} 2a_j b_j a_{k+1} b_{k+1} \leq \sum_{j=1}^{k} \left[\frac{\left(a_j^2 + b_j^2\right) a_{k+1}^2 b_{k+1}^2}{a_{k+1}^2 + b_{k+1}^2} + \frac{\left(a_{k+1}^2 + b_{k+1}^2\right) a_j^2 b_j^2}{a_j^2 + b_j^2}\right]$$

$$2a_{k+1} b_{k+1} \sum_{j=1}^{k} a_j b_j \leq \frac{a_{k+1}^2 b_{k+1}^2}{a_{k+1}^2 + b_{k+1}^2} \sum_{j=1}^{k} \left(a_j^2 + b_j^2\right) + \left(a_{k+1}^2 + b_{k+1}^2\right) \sum_{j=1}^{k} \frac{a_j^2 b_j^2}{a_j^2 + b_j^2}.$$

$$(72.3)$$

Adding (72.2) and (72.3) and then adding $a_{k+1}^2 b_{k+1}^2$ to both sides, we have the inequality

$$\left(\sum_{j=1}^{k+1} a_j b_j\right)^2 \leq \left(\sum_{j=1}^{k+1} \left(a_j^2 + b_j^2\right)\right) \left(\sum_{j=1}^{k+1} \frac{a_j^2 b_j^2}{a_j^2 + b_j^2}\right).$$

Thus, we proved that it is true for $n = k+1$ if it is true for $n = m$ for all $1 \leq m \leq k$.

Conclusion

We proved that it is true for $n = 1$ and $n = 2$. Then, we proved that it is true for $n = k+1$ if it is true for $n = m$ for all $1 \leq m \leq k$. Therefore, by strong induction,

$$\left(\sum_{j=1}^{n} a_j b_j\right)^2 \leq \left(\sum_{j=1}^{n} \left(a_j^2 + b_j^2\right)\right) \left(\sum_{j=1}^{n} \frac{a_j^2 b_j^2}{a_j^2 + b_j^2}\right)$$

for all $n \in \mathbb{N}$.

Problem 73

Topic: Binomial Coefficient

Problem: Prove the binomial theorem,

$$(x+y)^n = \sum_{j=0}^{n} \binom{n}{j} x^{n-j} y^j,$$

for all $n \in \mathbb{N}$.

Definition - Binomial Coefficient: For non-negative integers p and r,

$$\binom{p}{r} = \begin{cases} \frac{p!}{r!(p-r)!} & \text{if } r \le p \\ 0 & \text{if } r > p \end{cases}.$$

Solution

Base Cases

We need to prove that it is true for $n = 1$. We can check that

$$\sum_{j=0}^{1} \binom{1}{j} x^{1-j} y^j = \binom{1}{0} x^1 y^0 + \binom{1}{1} x^0 y^1 = x + y = (x+y)^1,$$

so it is true for $n = 1$.

Induction Step

Assume

$$(x + y)^k = \sum_{j=0}^{k} \binom{k}{j} x^{k-j} y^j \tag{73.1}$$

for some $k \geq 1$. To prove the induction step, we need to prove

$$(x + y)^{k+1} = \sum_{j=0}^{k+1} \binom{k+1}{j} x^{k+1-j} y^j.$$

Multiplying both sides of (73.1) by $x + y$,

$$(x + y)^{k+1} = \sum_{j=0}^{k} \binom{k}{j} x^{k+1-j} y^j + \sum_{j=0}^{k} \binom{k}{j} x^{k-j} y^{j+1}$$

$$= x^{k+1} + \sum_{j=1}^{k} \binom{k}{j} x^{k+1-j} y^j + \sum_{j=0}^{k-1} \binom{k}{j} x^{k-j} y^{j+1} + y^{k+1}.$$

Let $j + 1 \rightarrow j$,

$$\sum_{j=0}^{k-1} \binom{k}{j} x^{k-j} y^{j+1} = \sum_{j=1}^{k} \binom{k}{j-1} x^{k-(j-1)} y^j = \sum_{j=1}^{k} \binom{k}{j-1} x^{k+1-j} y^j.$$

So,

$$(x + y)^{k+1} = x^{k+1} + \sum_{j=1}^{k} \left[\binom{k}{j} + \binom{k}{j-1} \right] x^{k+1-j} y^j + y^{k+1}.$$

Since

$$\binom{k}{j} + \binom{k}{j-1} = \frac{k!}{j!(k-j)!} + \frac{k!}{(j-1)!(k-j+1)!}$$
$$= \frac{(k-j+1) \cdot k!}{j!(k-j+1)!} + \frac{j \cdot k!}{j!(k-j+1)!}$$
$$= \frac{(k+1)!}{j!(k+1-j)!} = \binom{k+1}{j},$$

which is known as Pascal's identity, we have

$$(x + y)^{k+1} = x^{k+1} + \sum_{j=1}^{k} \binom{k+1}{j} x^{k+1-j} y^j + y^{k+1}.$$

Note that

$$x^{k+1} = \binom{k+1}{0} x^{k+1-0} y^0$$

and

$$y^{k+1} = \binom{k+1}{k+1} x^{k+1-(k+1)} y^{k+1}.$$

Hence,

$$(x + y)^{k+1} = \sum_{j=0}^{k+1} \binom{k+1}{j} x^{k+1-j} y^j.$$

Thus, we proved that it is true for $n = k + 1$ if it is true for $n = k$.

Conclusion

We proved that it is true for $n = 1$. Then, we proved that it is true for $n = k + 1$ if it is true for $n = k$. Therefore, by weak induction,

$$(x + y)^n = \sum_{j=0}^{n} \binom{n}{j} x^{n-j} y^j$$

for all $n \in \mathbb{N}$.

Problem 74

Topic: Binomial Coefficient

Problem: For $n, r \in \mathbb{N}$, prove the Hockey-Stick identity,

$$\sum_{j=r}^{n} \binom{j}{r} = \binom{n+1}{r+1},$$

for $n \geq r$.

Definition - Binomial Coefficient: For non-negative integers p and r,

$$\binom{p}{r} = \begin{cases} \frac{p!}{r!(p-r)!} & \text{if } r \leq p \\ 0 & \text{if } r > p \end{cases}.$$

Solution

Base Cases

We need to prove that it is true for $n = r$. We can check that

$$\sum_{j=r}^{r} \binom{j}{r} = \binom{r}{r} = 1 = \binom{r+1}{r+1},$$

so it is true for $n = r$.

Induction Step

Assume

$$\sum_{j=r}^{k} \binom{j}{r} = \binom{k+1}{r+1} \qquad (74.1)$$

for some $k \geq r$. To prove the induction step, we need to prove

$$\sum_{j=r}^{k+1} \binom{j}{r} = \binom{k+2}{r+1}.$$

Using assumption (74.1),

$$\sum_{j=r}^{k+1} \binom{j}{r} = \sum_{j=r}^{k} \binom{j}{r} + \binom{k+1}{r}$$
$$= \binom{k+1}{r+1} + \binom{k+1}{r}.$$

In problem 73, we have calculated that

$$\binom{k}{j} + \binom{k}{j-1} = \binom{k+1}{j}.$$

Let $k \to k+1$ and $j \to r+1$,

$$\binom{k+1}{r+1} + \binom{k+1}{r} = \binom{k+2}{r+1}.$$

So,

$$\sum_{j=r}^{k+1} \binom{j}{r} = \binom{k+2}{r+1}.$$

Thus, we proved that it is true for $n = k+1$ if it is true for $n = k$.

Conclusion

We proved that it is true for $n = r$. Then, we proved that it is true for $n = k+1$ if it is true for $n = k$. Therefore, by weak induction,

$$\sum_{j=r}^{n} \binom{j}{r} = \binom{n+1}{r+1}$$

for $n \geq r$.

Problem 75

Topic: Binomial Coefficient, Inequality

Problem: Prove
$$\binom{2n}{n} \leq 2^{2n-1}$$
for all $n \in \mathbb{N}$.

Definition - Binomial Coefficient: For non-negative integers p and r,
$$\binom{p}{r} = \begin{cases} \frac{p!}{r!(p-r)!} & \text{if } r \leq p \\ 0 & \text{if } r > p \end{cases}.$$

Solution

Base Cases

We need to prove that it is true for $n = 1$. We can check that
$$\binom{2}{1} = \frac{2!}{1!\,1!} = 2 \leq 2 = 2^{2 \cdot 1 - 1},$$
so it is true for $n = 1$.

Induction Step

Assume
$$\binom{2k}{k} \leq 2^{2k-1} \tag{75.1}$$

for some $k \geq 1$. To prove the induction step, we need to prove

$$\binom{2k+2}{k+1} \leq 2^{2(k+1)-1} = 2^{2k+1}.$$

Note that

$$
\begin{aligned}
\binom{2k+2}{k+1} &= \frac{(2k+2)!}{((k+1)!)^2} \\
&= \frac{(2k+2)(2k+1)}{(k+1)^2} \cdot \frac{(2k)!}{(k!)^2} \\
&= \frac{(2k+2)(2k+1)}{(k+1)^2} \binom{2k}{k}.
\end{aligned}
$$

Since $2k+1 \leq 2k+2$, using assumption (75.1),

$$
\begin{aligned}
\binom{2k+2}{k+1} &\leq \frac{(2k+2)^2}{(k+1)^2} \binom{2k}{k} \\
&= 2^2 \binom{2k}{k} \\
&\leq 2^2 \cdot 2^{2k-1} \\
&= 2^{2k+1}.
\end{aligned}
$$

Thus, we proved that it is true for $n = k+1$ if it is true for $n = k$.

Conclusion

We proved that it is true for $n = 1$. Then, we proved that it is true for $n = k+1$ if it is true for $n = k$. Therefore, by weak induction,

$$\binom{2n}{n} \leq 2^{2n-1}$$

for all $n \in \mathbb{N}$.

Problem 76

Topic: Binomial Coefficient, Fibonacci Number

Problem: For any non-negative integer ℓ, prove

$$\sum_{j=0}^{n} \binom{n}{j} F_{\ell+j} = F_{\ell+2n}$$

for all $n \in \mathbb{N}$ where F_n denotes the n-th Fibonacci number.

Definition - Binomial Coefficient: For non-negative integers p and r,

$$\binom{p}{r} = \begin{cases} \frac{p!}{r!(p-r)!} & \text{if } r \le p \\ 0 & \text{if } r > p \end{cases}.$$

Definition - Fibonacci Numbers: The Fibonacci numbers are a sequence of numbers defined as $F_0 = 0$, $F_1 = 1$, and $F_n = F_{n-1} + F_{n-2}$ for $n \ge 2$.

Solution

Base Cases

We need to prove that it is true for $n = 1$. Since

$$\sum_{j=0}^{1} \binom{1}{j} F_{\ell+j} = \binom{1}{0} F_{\ell+0} + \binom{1}{1} F_{\ell+1} = F_\ell + F_{\ell+1} = F_{\ell+2},$$

it is true for $n = 1$.

Induction Step

Assume

$$\sum_{j=0}^{k}\binom{k}{j}F_{\ell+j} = F_{\ell+2k} \tag{76.1}$$

for some $k \geq 1$. To prove the induction step, we need to prove

$$\sum_{j=0}^{k+1}\binom{k+1}{j}F_{\ell+j} = F_{\ell+2(k+1)} = F_{\ell+2k+2}.$$

First, note that

$$\sum_{j=0}^{k+1}\binom{k+1}{j}F_{\ell+j} = \binom{k+1}{0}F_{\ell+0} + \sum_{j=1}^{k}\binom{k+1}{j}F_{\ell+j} + \binom{k+1}{k+1}F_{\ell+k+1}$$

$$= F_\ell + \sum_{j=1}^{k}\binom{k+1}{j}F_{\ell+j} + F_{\ell+k+1}.$$

In problem 73, we have calculated that

$$\binom{k+1}{j} = \binom{k}{j} + \binom{k}{j-1}.$$

So,

$$\sum_{j=0}^{k+1}\binom{k+1}{j}F_{\ell+j} = F_\ell + \sum_{j=1}^{k}\left[\binom{k}{j} + \binom{k}{j-1}\right]F_{\ell+j} + F_{\ell+k+1}$$

$$= F_\ell + \sum_{j=1}^{k}\binom{k}{j}F_{\ell+j} + \sum_{j=1}^{k}\binom{k}{j-1}F_{\ell+j} + F_{\ell+k+1}.$$

Substituting $j-1 \to j$,

$$\sum_{j=1}^{k}\binom{k}{j-1}F_{\ell+j} = \sum_{j=0}^{k-1}\binom{k}{j}F_{\ell+j+1}.$$

So,

$$\sum_{j=0}^{k+1}\binom{k+1}{j}F_{\ell+j} = F_\ell + \sum_{j=1}^{k}\binom{k}{j}F_{\ell+j} + \sum_{j=0}^{k-1}\binom{k}{j}F_{\ell+1+j} + F_{\ell+k+1}.$$

Note that

$$F_\ell + \sum_{j=1}^{k}\binom{k}{j}F_{\ell+j} = \binom{k}{0}F_{\ell+0} + \sum_{j=1}^{k}\binom{k}{j}F_{\ell+j} = \sum_{j=0}^{k}\binom{k}{j}F_{\ell+j}$$

and

$$\sum_{j=0}^{k-1}\binom{k}{j}F_{\ell+1+j} + F_{\ell+k+1} = \sum_{j=0}^{k-1}\binom{k}{j}F_{\ell+1+j} + \binom{k}{k}F_{\ell+1+k} = \sum_{j=0}^{k}\binom{k}{j}F_{\ell+1+j}.$$

Hence,
$$\sum_{j=0}^{k+1} \binom{k+1}{j} F_{\ell+j} = \sum_{j=0}^{k} \binom{k}{j} F_{\ell+j} + \sum_{j=0}^{k} \binom{k}{j} F_{\ell+1+j}.$$

By assumption (76.1),
$$\sum_{j=0}^{k} \binom{k}{j} F_{\ell+j} = F_{\ell+2k}$$

and
$$\sum_{j=0}^{k} \binom{k}{j} F_{\ell+1+j} = F_{\ell+1+2k},$$

so
$$\sum_{j=0}^{k+1} \binom{k+1}{j} F_{\ell+j} = F_{\ell+2k} + F_{\ell+1+2k} = F_{\ell+2k+2}$$

by using the recursive relation of F_n. Thus, we proved that it is true for $n = k+1$ if it is true for $n = k$.

Conclusion

We proved that it is true for $n = 1$. Then, we proved that it is true for $n = k+1$ if it is true for $n = k$. Therefore, by weak induction,

$$\sum_{j=0}^{n} \binom{n}{j} F_{\ell+j} = F_{\ell+2n}$$

for all $n \in \mathbb{N}$.

Problem 77

Topic: Binomial Coefficient, Fibonacci Number

Problem: Prove

$$\sum_{j=0}^{n} \binom{n-j}{j} = F_{n+1}$$

for all $n \in \mathbb{N}$ where F_n denotes the n-th Fibonacci number.

Definition - Binomial Coefficient: For non-negative integers p and r,

$$\binom{p}{r} = \begin{cases} \frac{p!}{r!(p-r)!} & \text{if } r \leq p \\ 0 & \text{if } r > p \end{cases}.$$

Definition - Fibonacci Numbers: The Fibonacci numbers are a sequence of numbers defined as $F_0 = 0$, $F_1 = 1$, and $F_n = F_{n-1} + F_{n-2}$ for $n \geq 2$.

Solution

Base Cases

We need to prove that it is true for $n = 1$ and $n = 2$. When $n = 1$,

$$\sum_{j=0}^{1} \binom{1-j}{j} = \binom{1}{0} + \binom{0}{1} = 1 + 0 = 1 = F_2$$

because $F_2 = F_1 + F_0 = 1 + 0 = 1$. When $n = 2$,

$$\sum_{j=0}^{2} \binom{2-j}{j} = \binom{2}{0} + \binom{1}{1} + \binom{0}{2} = 1 + 1 + 0 = 2 = F_3$$

because $F_3 = F_2 + F_1 = 1 + 1 = 2$. So, it is true for $n = 1$ and $n = 2$.

Induction Step

Assume

$$\sum_{j=0}^{m} \binom{m-j}{j} = F_{m+1} \tag{77.1}$$

for all $1 \le m \le k$ for some $k \ge 2$. To prove the induction step, we need to prove

$$\sum_{j=0}^{k+1} \binom{k+1-j}{j} = F_{k+2}.$$

Note that

$$\sum_{j=0}^{k+1} \binom{k+1-j}{j} = \binom{k+1}{0} + \sum_{j=1}^{k} \binom{k+1-j}{j} + \binom{0}{k+1}$$

$$= 1 + \sum_{j=1}^{k} \binom{k+1-j}{j}.$$

In problem 73, we have calculated that

$$\binom{k+1}{j} = \binom{k}{j} + \binom{k}{j-1}.$$

Let $k \to k - j$,

$$\binom{k+1-j}{j} = \binom{k-j}{j} + \binom{k-j}{j-1}.$$

So,

$$\sum_{j=0}^{k+1} \binom{k+1-j}{j} = 1 + \sum_{j=1}^{k} \binom{k-j}{j} + \sum_{j=1}^{k} \binom{k-j}{j-1}.$$

Note that

$$1 + \sum_{j=1}^{k} \binom{k-j}{j} = \binom{k-0}{0} + \sum_{j=1}^{k} \binom{k-j}{j} = \sum_{j=0}^{k} \binom{k-j}{j}.$$

Substituting $j - 1 \to j$,

$$\sum_{j=1}^{k} \binom{k-j}{j-1} = \sum_{j=0}^{k-1} \binom{k-1-j}{j}.$$

Hence,

$$\sum_{j=0}^{k+1} \binom{k+1-j}{j} = \sum_{j=0}^{k} \binom{k-j}{j} + \sum_{j=0}^{k-1} \binom{k-1-j}{j}.$$

By assumption (77.1) when $m = k$,

$$\sum_{j=0}^{k} \binom{k-j}{j} = F_{k+1}.$$

By assumption (77.1) when $m = k - 1$,

$$\sum_{j=0}^{k-1} \binom{k-1-j}{j} = F_k.$$

So,

$$\sum_{j=0}^{k+1} \binom{k+1-j}{j} = F_{k+1} + F_k = F_{k+2}$$

by using the recursive relation of F_n. Thus, we proved that it is true for $n = k+1$ if it is true for $n = m$ for all $1 \leq m \leq k$.

Conclusion

We proved that it is true for $n = 1$ and $n = 2$. Then, we proved that it is true for $n = k+1$ if it is true for $n = m$ for all $1 \leq m \leq k$. Therefore, by strong induction,

$$\sum_{j=0}^{n} \binom{n-j}{j} = F_{n+1}$$

for all $n \in \mathbb{N}$.

Problem 78

Topic: Derivative

Problem: Prove
$$\frac{d^n}{dx^n} x^n = n!$$
for all $n \in \mathbb{N}$.

Definition - Factorial:
$$n! = n \cdot (n-1) \cdot (n-2) \cdots 2 \cdot 1.$$
$$0! = 1.$$

Solution

Base Cases

We need to prove that it is true for $n = 1$. We can check that

$$\frac{d}{dx} x = 1 = 1!,$$

so it is true for $n = 1$.

Induction Step

Assume

$$\frac{d^k}{dx^k} x^k = k! \tag{78.1}$$

for some $k \geq 1$. To prove the induction step, we need to prove

$$\frac{d^{k+1}}{dx^{k+1}} x^{k+1} = (k+1)!.$$

Using assumption (78.1),

$$
\begin{aligned}
\frac{d^{k+1}}{dx^{k+1}} x^{k+1} &= \frac{d^k}{dx^k} \left(\frac{d}{dx} x^{k+1} \right) \\
&= \frac{d^k}{dx^k} \left((k+1)x^k \right) \\
&= (k+1) \frac{d^k}{dx^k} x^k \\
&= (k+1) \cdot k! \\
&= (k+1)!.
\end{aligned}
$$

Thus, we proved that it is true for $n = k+1$ if it is true for $n = k$.

Conclusion

We proved that it is true for $n = 1$. Then, we proved that it is true for $n = k+1$ if it is true for $n = k$. Therefore, by weak induction,

$$\frac{d^n}{dx^n} x^n = n!$$

for all $n \in \mathbb{N}$.

Problem 79

Topic: Derivative

> **Problem:** If $f(x)$ is an even function, then prove that $f^{(2n)}(x)$ is even for all $n \in \mathbb{N}$ where $f^{(n)}(x)$ denotes the n-th derivative of $f(x)$.

Solution

Base Cases

We need to prove that it is true for $n = 1$. If $f(x)$ is even, then

$$f(x) = f(-x).$$

Differentiating both sides with respect to x twice,

$$f^{(1)}(x) = -f^{(1)}(-x)$$
$$f^{(2)}(x) = -\left(-f^{(2)}(-x)\right) = f^{(2)}(-x).$$

So, $f^{(2)}(x)$ is even.

Induction Step

Assume that $f^{(2k)}(x)$ is even if $f(x)$ is even for some $k \geq 1$. To prove the induction step, we need to prove that $f^{(2k+2)}(x)$ is even. By the assumption that $f^{(2k)}(x)$ even,

$$f^{(2k)}(x) = f^{(2k)}(-x).$$

Differentiating both sides with respect to x twice,

$$f^{(2k+1)}(x) = -f^{(2k+1)}(-x)$$
$$f^{(2k+2)}(x) = -\left(-f^{(2k+2)}(-x)\right) = f^{(2k+2)}(-x).$$

Hence, $f^{(2k+2)}(x)$ is even. Thus, we proved that it is true for $n = k+1$ if it is true for $n = k$.

Conclusion

We proved that it is true for $n = 1$. Then, we proved that it is true for $n = k+1$ if it is true for $n = k$. Therefore, by weak induction, if $f(x)$ is an even function, then $f^{(2n)}(x)$ is even for all $n \in \mathbb{N}$.

Problem 80

Topic: Derivative, Trigonometry

Problem: Prove
$$\frac{d^n}{dx^n} \sin x = \sin\left(x + \frac{n\pi}{2}\right)$$
for all $n \in \mathbb{N}$.

Solution

Base Cases

We need to prove that it is true for $n = 1$. Since
$$\frac{d}{dx} \sin x = \cos x$$

and
$$\sin\left(x + \frac{\pi}{2}\right) = \sin x \cos \frac{\pi}{2} + \cos x \sin \frac{\pi}{2} = \cos x,$$

it is true that
$$\frac{d}{dx} \sin x = \sin\left(x + \frac{\pi}{2}\right).$$

Induction Step

Assume
$$\frac{d^k}{dx^k} \sin x = \sin\left(x + \frac{k\pi}{2}\right) \tag{80.1}$$

for some $k \geq 1$. To prove the induction step, we need to prove
$$\frac{d^{k+1}}{dx^{k+1}} \sin x = \sin\left(x + \frac{(k+1)\pi}{2}\right).$$

Differentiating both sides of (80.1) with respect to x using the chain rule,

$$\frac{d^{k+1}}{dx^{k+1}} \sin x = \frac{d}{dx} \sin\left(x + \frac{k\pi}{2}\right) = \cos\left(x + \frac{k\pi}{2}\right).$$

By using the addition formula of sine,

$$\cos\left(x + \frac{k\pi}{2}\right) = \sin\left(x + \frac{k\pi}{2}\right)\cos\left(\frac{\pi}{2}\right) + \cos\left(x + \frac{k\pi}{2}\right)\sin\left(\frac{\pi}{2}\right)$$

$$= \sin\left(x + \frac{(k+1)\pi}{2}\right).$$

Hence,

$$\frac{d^{k+1}}{dx^{k+1}} \sin x = \sin\left(x + \frac{(k+1)\pi}{2}\right).$$

Thus, we proved that it is true for $n = k + 1$ if it is true for $n = k$.

Conclusion

We proved that it is true for $n = 1$. Then, we proved that it is true for $n = k + 1$ if it is true for $n = k$. Therefore, by weak induction,

$$\frac{d^n}{dx^n} \sin x = \sin\left(x + \frac{n\pi}{2}\right)$$

for all $n \in \mathbb{N}$.

d
—
dx

Problem 81

Topic: Derivative, Trigonometry

Problem: Prove
$$\frac{d^n}{dx^n} \cos x = \cos\left(x + \frac{n\pi}{2}\right)$$
for all $n \in \mathbb{N}$.

Solution

Base Cases

We need to prove that it is true for $n = 1$. Since
$$\frac{d}{dx} \cos x = -\sin x$$
and
$$\cos\left(x + \frac{\pi}{2}\right) = \cos x \cos \frac{\pi}{2} - \sin x \sin \frac{\pi}{2} = -\sin x,$$
is true that
$$\frac{d}{dx} \cos x = \cos\left(x + \frac{\pi}{2}\right).$$

Induction Step

Assume
$$\frac{d^k}{dx^k} \cos x = \cos\left(x + \frac{k\pi}{2}\right) \tag{81.1}$$
for some $k \geq 1$. To prove the induction step, we need to prove
$$\frac{d^{k+1}}{dx^{k+1}} \cos x = \cos\left(x + \frac{(k+1)\pi}{2}\right).$$

Differentiating both sides of (81.1) with respect to x using the chain rule,

$$\frac{d^{k+1}}{dx^{k+1}} \cos x = \frac{d}{dx} \cos\left(x + \frac{k\pi}{2}\right) = -\sin\left(x + \frac{k\pi}{2}\right).$$

By using the addition formula of cosine,

$$-\sin\left(x + \frac{k\pi}{2}\right) = \cos\left(x + \frac{k\pi}{2}\right)\cos\left(\frac{\pi}{2}\right) - \sin\left(x + \frac{k\pi}{2}\right)\sin\left(\frac{\pi}{2}\right)$$

$$= \cos\left(x + \frac{(k+1)\pi}{2}\right).$$

Hence,

$$\frac{d^{k+1}}{dx^{k+1}} \cos x = \cos\left(x + \frac{(k+1)\pi}{2}\right).$$

Thus, we proved that it is true for $n = k + 1$ if it is true for $n = k$.

Conclusion

We proved that it is true for $n = 1$. Then, we proved that it is true for $n = k + 1$ if it is true for $n = k$. Therefore, by weak induction,

$$\frac{d^n}{dx^n} \cos x = \cos\left(x + \frac{n\pi}{2}\right)$$

for all $n \in \mathbb{N}$.

Problem 82

Topic: Derivative

Problem: Prove
$$\frac{d^n}{dx^n}\ln x = \frac{(-1)^{n-1}(n-1)!}{x^n}$$
for all $n \in \mathbb{N}$.

Definition - Factorial:
$$n! = n \cdot (n-1) \cdot (n-2) \cdots 2 \cdot 1.$$
$$0! = 1.$$

Solution

Base Cases

We need to prove that it is true for $n = 1$. We can check that

$$\frac{d}{dx}\ln x = \frac{1}{x} = \frac{(-1)^{1-1}(1-1)!}{x^1},$$

so it is true for $n = 1$.

Induction Step

Assume

$$\frac{d^k}{dx^k}\ln x = \frac{(-1)^{k-1}(k-1)!}{x^k} \tag{82.1}$$

for some $k \geq 1$. To prove the induction step, we need to prove

$$\frac{d^{k+1}}{dx^{k+1}} \ln x = \frac{(-1)^k \, k!}{x^{k+1}}.$$

Differentiating both sides of (82.1) with respect to x,

$$\begin{aligned}
\frac{d^{k+1}}{dx^{k+1}} \ln x &= \frac{d}{dx} \frac{(-1)^{k-1}(k-1)!}{x^k} \\
&= (-1)^{k-1}(k-1)! \times \frac{d}{dx} \frac{1}{x^k} \\
&= (-1)^{k-1}(k-1)! \times \frac{-k}{x^{k+1}} \\
&= \frac{(-1)^k \, k!}{x^{k+1}}.
\end{aligned}$$

Thus, we proved that it is true for $n = k + 1$ if it is true for $n = k$.

Conclusion

We proved that it is true for $n = 1$. Then, we proved that it is true for $n = k + 1$ if it is true for $n = k$. Therefore, by weak induction,

$$\frac{d^n}{dx^n} \ln x = \frac{(-1)^{n-1}(n-1)!}{x^n}$$

for all $n \in \mathbb{N}$.

Problem 83

Topic: Derivative

Problem: For $a, b, c, d \in \mathbb{R}$ and $b, c \neq 0$, prove

$$\frac{d^n}{dx^n} \frac{ax+b}{cx+d} = \frac{n!(-c)^{n-1}(ad-bc)}{(cx+d)^{n+1}}$$

for all $n \in \mathbb{N}$.

Definition - Factorial:

$$n! = n \cdot (n-1) \cdot (n-2) \cdots 2 \cdot 1.$$

$$0! = 1.$$

Solution

Base Cases

We need to prove that it is true for $n = 1$. By the quotient rule,

$$\frac{d}{dx} \frac{ax+b}{cx+d} = \frac{a(cx+d) - c(ax+b)}{(cx+d)^2} = \frac{ad-bc}{(cx+d)^2} = \frac{1!(-c)^{1-1}(ad-bc)}{(cx+d)^{1+1}},$$

so it is true for $n = 1$.

Induction Step

Assume

$$\frac{d^k}{dx^k} \frac{ax+b}{cx+d} = \frac{k!(-c)^{k-1}(ad-bc)}{(cx+d)^{k+1}} \tag{83.1}$$

for some $k \geq 1$. To prove the induction step, we need to prove

$$\frac{d^{k+1}}{dx^{k+1}} \frac{ax+b}{cx+d} = \frac{(k+1)!(-c)^k(ad-bc)}{(cx+d)^{k+2}}.$$

Differentiating both sides of (83.1) with respect to x,

$$\frac{d^{k+1}}{dx^{k+1}} \frac{ax+b}{cx+d} = \frac{d}{dx} \frac{k!(-c)^{k-1}(ad-bc)}{(cx+d)^{k+1}}$$

$$= k!(-c)^{k-1}(ad-bc) \times \frac{d}{dx} \frac{1}{(cx+d)^{k+1}}$$

$$= k!(-c)^{k-1}(ad-bc) \times \frac{c(-k-1)}{(cx+d)^{k+2}}$$

$$= \frac{(k+1)!(-c)^k(ad-bc)}{(cx+d)^{k+2}}.$$

Thus, we proved that it is true for $n = k+1$ if it is true for $n = k$.

Conclusion

We proved that it is true for $n = 1$. Then, we proved that it is true for $n = k+1$ if it is true for $n = k$. Therefore, by weak induction,

$$\frac{d^n}{dx^n} \frac{ax+b}{cx+d} = \frac{n!(-c)^{n-1}(ad-bc)}{(cx+d)^{n+1}}$$

for all $n \in \mathbb{N}$.

Problem 84

Topic: Derivative

Problem: Prove

$$\frac{d^n}{dx^n}\frac{1}{\sqrt{1-x}} = \frac{(2n)!}{4^n\,n!}\cdot\frac{1}{(1-x)^{n+\frac{1}{2}}}$$

for all $n \in \mathbb{N}$.

Definition - Factorial:

$$n! = n\cdot(n-1)\cdot(n-2)\cdots 2\cdot 1.$$

$$0! = 1.$$

Solution

Base Cases

We need to prove that it is true for $n = 1$. We can check that

$$\frac{d}{dx}\frac{1}{\sqrt{1-x}} = \frac{1}{2(1-x)^{3/2}} = \frac{2!}{4^1\cdot 1!}\cdot\frac{1}{(1-x)^{1+\frac{1}{2}}},$$

it is true for $n = 1$.

Induction Step

Assume

$$\frac{d^k}{dx^k}\frac{1}{\sqrt{1-x}} = \frac{(2k)!}{4^k\,k!}\cdot\frac{1}{(1-x)^{k+\frac{1}{2}}} \tag{84.1}$$

for some $k \geq 1$. To prove the induction step, we need to prove

$$\frac{d^{k+1}}{dx^{k+1}} \frac{1}{\sqrt{1-x}} = \frac{(2k+2)!}{4^{k+1}(k+1)!} \cdot \frac{1}{(1-x)^{k+1+\frac{1}{2}}}.$$

Differentiating both sides of (84.1) with respect to x,

$$\begin{aligned}
\frac{d^{k+1}}{dx^{k+1}} \frac{1}{\sqrt{1-x}} &= \frac{d}{dx} \frac{(2k)!}{4^k\, k!} \cdot \frac{1}{(1-x)^{k+\frac{1}{2}}} \\
&= \frac{(2k)!}{4^k\, k!} \times \frac{d}{dx} \frac{1}{(1-x)^{k+\frac{1}{2}}} \\
&= \frac{(2k)!}{4^k\, k!} \times \frac{k+\frac{1}{2}}{(1-x)^{k+\frac{1}{2}+1}} \\
&= \frac{(2k)!}{4^k\, k!} \times \frac{(2k+1)(2k+2)}{4(k+1)} \times \frac{1}{(1-x)^{k+1+\frac{1}{2}}} \\
&= \frac{(2k+2)!}{4^{k+1}(k+1)!} \cdot \frac{1}{(1-x)^{k+1+\frac{1}{2}}}.
\end{aligned}$$

Thus, we proved that it is true for $n = k + 1$ if it is true for $n = k$.

Conclusion

We proved that it is true for $n = 1$. Then, we proved that it is true for $n = k +$ if it is true for $n = k$. Therefore, by weak induction,

$$\frac{d^n}{dx^n} \frac{1}{\sqrt{1-x}} = \frac{(2n)!}{4^n\, n!} \cdot \frac{1}{(1-x)^{n+\frac{1}{2}}}$$

for all $n \in \mathbb{N}$.

Problem 85

Topic: Derivative, Trigonometry

Problem: Prove

$$\frac{d^n}{dx^n} \tan^{-1} x = \frac{(-1)^{n-1}(n-1)!}{(1+x^2)^{n/2}} \sin\left(n \cot^{-1} x\right)$$

for all $n \in \mathbb{N}$.

Definition - Factorial:

$$n! = n \cdot (n-1) \cdot (n-2) \cdots 2 \cdot 1.$$

$$0! = 1.$$

Solution

Base Cases

We need to prove that it is true for $n = 1$. Since

$$\sin^2\left(\cot^{-1} x\right) + \cos^2\left(\cot^{-1} x\right) = 1$$

and

$$\frac{\cos\left(\cot^{-1} x\right)}{\sin\left(\cot^{-1} x\right)} = \cot\left(\cot^{-1} x\right) = x,$$

we have

$$\sin\left(\cot^{-1} x\right) = \frac{1}{\sqrt{1+x^2}}.$$

So,

$$\frac{d}{dx}\tan^{-1}x = \frac{1}{1+x^2}$$

$$= \frac{1}{\sqrt{1+x^2}} \cdot \frac{1}{\sqrt{1+x^2}}$$

$$= \frac{(-1)^{1-1}(1-1)!}{(1+x^2)^{1/2}}\sin\left(\cot^{-1}x\right).$$

Induction Step

Assume

$$\frac{d^k}{dx^k}\tan^{-1}x = \frac{(-1)^{k-1}(k-1)!}{(1+x^2)^{k/2}}\sin\left(k\cot^{-1}x\right) \tag{85.1}$$

for some $k \geq 1$. To prove the induction step, we need to prove

$$\frac{d^{k+1}}{dx^{k+1}}\tan^{-1}x = \frac{(-1)^k\,k!}{(1+x^2)^{(k+1)/2}}\sin\left((k+1)\cot^{-1}x\right).$$

Differentiating both sides of (85.1) with respect to x using the product rule,

$$\frac{d^{k+1}}{dx^{k+1}}\tan^{-1}x = \frac{d}{dx}\frac{(-1)^{k-1}(k-1)!}{(1+x^2)^{k/2}}\sin\left(k\cot^{-1}x\right)$$

$$= (-1)^{k-1}(k-1)!\sin\left(k\cot^{-1}x\right)\times\frac{d}{dx}\frac{1}{(1+x^2)^{k/2}}$$

$$+ \frac{(-1)^{k-1}(k-1)!}{(1+x^2)^{k/2}}\times\frac{d}{dx}\sin\left(k\cot^{-1}x\right).$$

Using the chain rule,

$$\frac{d}{dx}\frac{1}{(1+x^2)^{k/2}} = \frac{-kx}{(1+x^2)^{(k+2)/2}}$$

and

$$\frac{d}{dx}\sin\left(k\cot^{-1}x\right) = \cos\left(k\cot^{-1}x\right)\cdot\frac{-k}{1+x^2}.$$

So,

$$\frac{d^{k+1}}{dx^{k+1}}\tan^{-1}x = (-1)^{k-1}(k-1)!\sin\left(k\cot^{-1}x\right)\times\frac{-kx}{(1+x^2)^{(k+2)/2}}$$

$$+ \frac{(-1)^{k-1}(k-1)!}{(1+x^2)^{k/2}}\times\frac{-k\cos\left(k\cot^{-1}x\right)}{1+x^2}$$

$$= \frac{(-1)^k\,k!}{(1+x^2)^{(k+1)/2}}\left[\frac{x}{\sqrt{1+x^2}}\sin\left(k\cot^{-1}x\right)\right.$$

$$\left.+ \frac{1}{\sqrt{1+x^2}}\cos\left(k\cot^{-1}x\right)\right].$$

Since

$$\sin^2\left(\cot^{-1} x\right) + \cos^2\left(\cot^{-1} x\right) = 1$$

and

$$\frac{\cos\left(\cot^{-1} x\right)}{\sin\left(\cot^{-1} x\right)} = \cot\left(\cot^{-1} x\right) = x,$$

we have

$$\sin\left(\cot^{-1} x\right) = \frac{1}{\sqrt{1+x^2}}$$

and

$$\cos\left(\cot^{-1} x\right) = \frac{x}{\sqrt{1+x^2}}.$$

Hence,

$$\begin{aligned}
\frac{d^{k+1}}{dx^{k+1}} \tan^{-1} x &= \frac{(-1)^k\, k!}{\left(1+x^2\right)^{(k+1)/2}}\left[\cos\left(\cot^{-1} x\right)\sin\left(k\,\cot^{-1} x\right)\right.\\
&\qquad\left. + \sin\left(\cot^{-1} x\right)\cos\left(k\,\cot^{-1} x\right)\right]\\
&= \frac{(-1)^k\, k!}{\left(1+x^2\right)^{(k+1)/2}}\sin\left((k+1)\cot^{-1} x\right).
\end{aligned}$$

Thus, we proved that it is true for $n = k+1$ if it is true for $n = k$.

Conclusion

We proved that it is true for $n = 1$. Then, we proved that it is true for $n = k+1$ if it is true for $n = k$. Therefore, by weak induction,

$$\frac{d^n}{dx^n} \tan^{-1} x = \frac{(-1)^{n-1}(n-1)!}{\left(1+x^2\right)^{n/2}}\sin\left(n\,\cot^{-1} x\right)$$

for all $n \in \mathbb{N}$.

Problem 86

Topic: Derivative, Hyperbolic Functions

Problem: Prove

$$\frac{d^{2n}}{dx^{2n}} \tanh^{-1} x = \frac{(2n-1)!}{(1-x^2)^n} \sinh\left(2n \, \tanh^{-1} x\right)$$

for all $n \in \mathbb{N}$.

Definition - Factorial:

$$n! = n \cdot (n-1) \cdot (n-2) \cdots 2 \cdot 1.$$

$$0! = 1.$$

Solution

Base Cases

We need to prove that it is true for $n = 1$. The second derivative of $\tanh^{-1} x$ is

$$\frac{d^2}{dx^2} \tanh^{-1} x = \frac{d}{dx} \frac{1}{1-x^2} = \frac{2x}{(1-x^2)^2}.$$

The hyperbolic cosine $\cosh x$ and the hyperbolic sine $\sinh x$ satisfy the identity

$$\cosh^2 x - \sinh^2 x = 1,$$

which can be proven using the definitions

$$\cosh x = \frac{e^x + e^{-x}}{2}$$

and

$$\sinh x = \frac{e^x - e^{-x}}{2}.$$

Since

$$\cosh^2\left(\tanh^{-1} x\right) - \sinh^2\left(\tanh^{-1} x\right) = 1$$

and

$$\frac{\sinh\left(\tanh^{-1} x\right)}{\cosh\left(\tanh^{-1} x\right)} = \tanh\left(\tanh^{-1} x\right) = x,$$

we have

$$\sinh\left(\tanh^{-1} x\right) = \frac{x}{\sqrt{1-x^2}}$$

and

$$\cosh\left(\tanh^{-1} x\right) = \frac{1}{\sqrt{1-x^2}}.$$

Using the identity $\sinh(2x) = 2\sinh x \cosh x$,

$$\sinh\left(2\tanh^{-1} x\right) = 2\sinh\left(\tanh^{-1} x\right)\cosh\left(\tanh^{-1} x\right)$$

$$= 2 \cdot \frac{x}{\sqrt{1-x^2}} \cdot \frac{1}{\sqrt{1-x^2}} = \frac{2x}{1-x^2}.$$

So,

$$\frac{(2\cdot 1 - 1)!}{\left(1-x^2\right)^1}\sinh\left(2\tanh^{-1} x\right) = \frac{1}{1-x^2}\cdot\frac{2x}{1-x^2} = \frac{2x}{\left(1-x^2\right)^2}.$$

Hence, it is true that

$$\frac{d^2}{dx^2}\tanh^{-1} x = \frac{(2\cdot 1 - 1)!}{\left(1-x^2\right)^1}\sinh\left(2\tanh^{-1} x\right).$$

Induction Step

Assume

$$\frac{d^{2k}}{dx^{2k}}\tanh^{-1} x = \frac{(2k-1)!}{\left(1-x^2\right)^k}\sinh\left(2k\tanh^{-1} x\right) \tag{86.1}$$

for some $k \geq 1$. To prove the induction step, we need to prove

$$\frac{d^{2k+2}}{dx^{2k+2}}\tanh^{-1} x = \frac{(2k+1)!}{\left(1-x^2\right)^{k+1}}\sinh\left((2k+2)\tanh^{-1} x\right).$$

Differentiating both sides of (86.1) with respect to x using the product rule,

$$\frac{d^{2k+1}}{dx^{2k+1}}\tanh^{-1} x = (2k-1)!\,\sinh\left(2k\tanh^{-1} x\right)\times\frac{d}{dx}\frac{1}{\left(1-x^2\right)^k}$$

$$+ \frac{(2k-1)!}{\left(1-x^2\right)^k}\times\frac{d}{dx}\sinh\left(2k\tanh^{-1} x\right).$$

Using the chain rule,

$$\frac{d}{dx} \frac{1}{(1-x^2)^k} = \frac{2kx}{(1-x^2)^{k+1}}$$

and

$$\frac{d}{dx} \sinh\left(2k \tanh^{-1} x\right) = \cosh\left(2k \tanh^{-1} x\right) \cdot \frac{2k}{1-x^2}.$$

Hence,

$$\frac{d^{2k+1}}{dx^{2k+1}} \tanh^{-1} x = (2k-1)! \sinh\left(2k \tanh^{-1} x\right) \times \frac{2kx}{(1-x^2)^{k+1}}$$

$$+ \frac{(2k-1)!}{(1-x^2)^k} \times \frac{2k \cosh\left(2k \tanh^{-1} x\right)}{1-x^2}$$

$$= \frac{(2k)!}{(1-x^2)^{k+\frac{1}{2}}} \left[\frac{x}{\sqrt{1-x^2}} \sinh\left(2k \tanh^{-1} x\right) \right.$$

$$\left. + \frac{1}{\sqrt{1-x^2}} \cosh\left(2k \tanh^{-1} x\right) \right]$$

$$= \frac{(2k)!}{(1-x^2)^{k+\frac{1}{2}}} \left[\sinh\left(\tanh^{-1} x\right) \sinh\left(2k \tanh^{-1} x\right) \right.$$

$$\left. + \cosh\left(\tanh^{-1} x\right) \cosh\left(2k \tanh^{-1} x\right) \right].$$

Using the identity

$$\cosh(x + y) = \cosh x \cosh y + \sinh x \sinh y,$$

we get

$$\frac{d^{2k+1}}{dx^{2k+1}} \tanh^{-1} x = \frac{(2k)!}{(1-x^2)^{k+\frac{1}{2}}} \cosh\left((2k+1) \tanh^{-1} x\right).$$

Differentiating both sides with respect to x again using the product rule,

$$\frac{d^{2k+2}}{dx^{2k+2}} \tanh^{-1} x = (2k)! \cosh\left((2k+1) \tanh^{-1} x\right) \times \frac{d}{dx} \frac{1}{(1-x^2)^{k+\frac{1}{2}}}$$

$$+ \frac{(2k)!}{(1-x^2)^{k+\frac{1}{2}}} \times \frac{d}{dx} \cosh\left((2k+1) \tanh^{-1} x\right).$$

Using the chain rule,

$$\frac{d}{dx} \frac{1}{(1-x^2)^{k+\frac{1}{2}}} = \frac{(2k+1)x}{(1-x^2)^{k+\frac{3}{2}}}$$

and

$$\frac{d}{dx} \cosh\left((2k+1) \tanh^{-1} x\right) = \sinh\left((2k+1) \tanh^{-1} x\right) \cdot \frac{2k+1}{1-x^2}.$$

Hence,

$$\frac{d^{2k+2}}{dx^{2k+2}} \tanh^{-1} x = (2k)! \cosh\left((2k+1)\tanh^{-1} x\right) \times \frac{(2k+1)x}{(1-x^2)^{k+\frac{3}{2}}}$$

$$+ \frac{(2k)!}{(1-x^2)^{k+\frac{1}{2}}} \times \frac{(2k+1)\sinh\left((2k+1)\tanh^{-1} x\right)}{1-x^2}$$

$$= \frac{(2k+1)!}{(1-x^2)^{k+1}} \left[\frac{x}{\sqrt{1-x^2}} \cosh\left((2k+1)\tanh^{-1} x\right) \right.$$

$$\left. + \frac{1}{\sqrt{1-x^2}} \sinh\left((2k+1)\tanh^{-1} x\right) \right]$$

$$= \frac{(2k+1)!}{(1-x^2)^{k+1}} \left[\sinh\left(\tanh^{-1} x\right) \cosh\left((2k+1)\tanh^{-1} x\right) \right.$$

$$\left. + \cosh\left(\tanh^{-1} x\right) \sinh\left((2k+1)\tanh^{-1} x\right) \right].$$

Using the identity

$$\sinh(x+y) = \sinh x \cosh y + \cosh x \sinh y,$$

we get

$$\frac{d^{2k+2}}{dx^{2k+2}} \tanh^{-1} x = \frac{(2k+1)!}{(1-x^2)^{k+1}} \sinh\left((2k+2)\tanh^{-1} x\right).$$

Thus, we proved that it is true for $n = k+1$ if it is true for $n = k$.

Conclusion

We proved that it is true for $n = 1$. Then, we proved that it is true for $n = k+1$ if it is true for $n = k$. Therefore, by weak induction,

$$\frac{d^{2n}}{dx^{2n}} \tanh^{-1} x = \frac{(2n-1)!}{(1-x^2)^n} \sinh\left(2n \tanh^{-1} x\right)$$

for all $n \in \mathbb{N}$.

Problem 87

Topic: Derivative, Binomial Coefficient

Problem: Prove the Leibniz rule,

$$\frac{d^n}{dx^n}(f(x)g(x)) = \sum_{j=0}^{n} \binom{n}{j} f^{(j)}(x) g^{(n-j)}(x),$$

for all $n \in \mathbb{N}$ where $f^{(n)}(x)$ denotes the n-th derivative of $f(x)$.

Definition - Binomial Coefficient: For non-negative integers p and r,

$$\binom{p}{r} = \begin{cases} \frac{p!}{r!(p-r)!} & \text{if } r \leq p \\ 0 & \text{if } r > p \end{cases}.$$

Solution

Base Cases

We need to prove that it is true for $n = 1$. By definition of derivative,

$$f'(x) = \lim_{h \to 0} \frac{f(x+h) - f(x)}{h}.$$

So,

$$\frac{d}{dx}(f(x)g(x)) = \lim_{h \to 0} \frac{f(x+h)g(x+h) - f(x)g(x)}{h}.$$

Adding and subtracting $f(x)g(x+h)$ in the numerator,

$$\frac{d}{dx}(f(x)g(x))$$

$$= \lim_{h \to 0} \frac{f(x+h)g(x+h) - f(x)g(x+h) + f(x)g(x+h) - f(x)g(x)}{h}$$

$$= \lim_{h \to 0} \frac{f(x+h)g(x+h) - f(x)g(x+h)}{h} + \lim_{h \to 0} \frac{f(x)g(x+h) - f(x)g(x)}{h}$$

$$= \left(\lim_{h \to 0} \frac{f(x+h) - f(x)}{h} \right) \cdot g(x) + f(x) \cdot \left(\lim_{h \to 0} \frac{g(x+h) - g(x)}{h} \right)$$

$$= f'(x)g(x) + f(x)g'(x).$$

For the right-hand side,

$$\sum_{j=0}^{1} \binom{1}{j} f^{(j)}(x)g^{(1-j)}(x) = \binom{1}{0} f^{(0)}(x)g^{(1)}(x) + \binom{1}{1} f^{(1)}(x)g^{(0)}(x)$$

$$= f'(x)g(x) + f(x)g'(x).$$

So, it is true that

$$\frac{d}{dx}(f(x)g(x)) = \sum_{j=0}^{1} \binom{1}{j} f^{(j)}(x)g^{(1-j)}(x).$$

Induction Step

Assume

$$\frac{d^m}{dx^m}(f(x)g(x)) = \sum_{j=0}^{m} \binom{m}{j} f^{(j)}(x)g^{(m-j)}(x) \tag{87.1}$$

for all $1 \leq m \leq k$ for some $k \geq 1$. To prove the induction step, we need to prove

$$\frac{d^{k+1}}{dx^{k+1}}(f(x)g(x)) = \sum_{j=0}^{k+1} \binom{k+1}{j} f^{(j)}(x)g^{(k+1-j)}(x).$$

By assumption (87.1) when $m = k$,

$$\frac{d^k}{dx^k}(f(x)g(x)) = \sum_{j=0}^{k} \binom{k}{j} f^{(j)}(x)g^{(k-j)}(x).$$

Differentiating both sides with respect to x,

$$\frac{d^{k+1}}{dx^{k+1}}(f(x)g(x)) = \frac{d}{dx} \sum_{j=0}^{k} \binom{k}{j} f^{(j)}(x)g^{(k-j)}(x)$$

$$= \sum_{j=0}^{k} \binom{k}{j} \frac{d}{dx} \left(f^{(j)}(x)g^{(k-j)}(x) \right).$$

By assumption (87.1) when $m = 1$,

$$\frac{d}{dx}(f(x)g(x)) = \sum_{j=0}^{1} \binom{1}{j} f^{(j)}(x)g^{(1-j)}(x) = f'(x)g(x) + f(x)g'(x).$$

So,

$$\frac{d}{dx}\left(f^{(j)}(x)g^{(k-j)}(x)\right) = f^{(j+1)}(x)g^{(k-j)}(x) + f^{(j)}(x)g^{(k+1-j)}(x),$$

and we have

$$\frac{d^{k+1}}{dx^{k+1}}(f(x)g(x)) = \sum_{j=0}^{k} \binom{k}{j} \left[f^{(j+1)}(x)g^{(k-j)}(x) + f^{(j)}(x)g^{(k+1-j)}(x)\right]$$

$$= \sum_{j=0}^{k} \binom{k}{j} f^{(j+1)}(x)g^{(k-j)}(x) + \sum_{j=0}^{k} \binom{k}{j} f^{(j)}(x)g^{(k+1-j)}(x).$$

Substituting $j + 1 \to j$,

$$\sum_{j=0}^{k} \binom{k}{j} f^{(j+1)}(x)g^{(k-j)}(x) = \sum_{j=1}^{k+1} \binom{k}{j-1} f^{(j)}(x)g^{(k+1-j)}(x).$$

Hence,

$$\frac{d^{k+1}}{dx^{k+1}}(f(x)g(x))$$

$$= \sum_{j=1}^{k+1} \binom{k}{j-1} f^{(j)}(x)g^{(k+1-j)}(x) + \sum_{j=0}^{k} \binom{k}{j} f^{(j)}(x)g^{(k+1-j)}(x)$$

$$= \binom{k}{k} f^{(k+1)}(x)g^{(k+1-k-1)}(x) + \sum_{j=1}^{k} \binom{k}{j-1} f^{(j)}(x)g^{(k+1-j)}(x)$$

$$+ \sum_{j=1}^{k} \binom{k}{j} f^{(j)}(x)g^{(k+1-j)}(x) + \binom{k}{0} f^{(0)}(x)g^{(k+1-0)}(x)$$

$$= f^{(k+1)}(x)g(x) + \sum_{j=1}^{k} \left[\binom{k}{j-1} + \binom{k}{j}\right] f^{(j)}(x)g^{(k+1-j)}(x) + f(x)g^{(k+1)}(x).$$

In problem 73, we have calculated that

$$\binom{k}{j-1} + \binom{k}{j} = \binom{k+1}{j}.$$

So,

$$\frac{d^{k+1}}{dx^{k+1}}\left(f(x)g(x)\right)$$

$$=f^{(k+1)}(x)g(x)+\sum_{j=1}^{k}\binom{k+1}{j}f^{(j)}(x)g^{(k+1-j)}(x)+f(x)g^{(k+1)}(x)$$

$$=\binom{k+1}{k+1}f^{(k+1)}(x)g^{(k+1-k-1)}(x)+\sum_{j=1}^{k}\binom{k+1}{j}f^{(j)}(x)g^{(k+1-j)}(x)$$

$$+\binom{k+1}{0}f^{(0)}(x)g^{(k+1-0)}(x)$$

$$=\sum_{j=0}^{k+1}\binom{k+1}{j}f^{(j)}(x)g^{(k+1-j)}(x).$$

Thus, we proved that it is true for $n=k+1$ if it is true for $n=m$ for all $1\leq m\leq k$.

Conclusion

We proved that it is true for $n=1$. Then, we proved that it is true for $n=k+1$ if it is true for $n=m$ for all $1\leq m\leq k$. Therefore, by strong induction,

$$\frac{d^n}{dx^n}\left(f(x)g(x)\right)=\sum_{j=0}^{n}\binom{n}{j}f^{(j)}(x)g^{(n-j)}(x)$$

for all $n\in\mathbb{N}$.

Problem 88

Topic: Derivative

Problem: Prove

$$\frac{d}{dx} \prod_{j=1}^{n} f_j(x) = \prod_{j=1}^{n} f_j(x) \cdot \left(\sum_{j=1}^{n} \frac{f_j'(x)}{f_j(x)} \right)$$

for all $n \in \mathbb{N}$.

Solution

Base Cases

We need to prove that it is true for $n = 1$ and $n = 2$. When $n = 1$,

$$\frac{d}{dx} \prod_{j=1}^{1} f_j(x) = f_1'(x) = f_1(x) \cdot \frac{f_1'(x)}{f_1(x)} = \prod_{j=1}^{1} f_j(x) \cdot \left(\sum_{j=1}^{1} \frac{f_j'(x)}{f_j(x)} \right).$$

When $n = 2$, by the product rule of derivative,

$$\frac{d}{dx} \prod_{j=1}^{2} f_j(x) = \frac{d}{dx} \left(f_1(x) f_2(x) \right) = f_1'(x) f_2(x) + f_1(x) f_2'(x)$$

$$= f_1(x) f_2(x) \cdot \left(\frac{f_1'(x)}{f_1(x)} + \frac{f_2'(x)}{f_2(x)} \right) = \prod_{j=1}^{2} f_j(x) \cdot \left(\sum_{j=1}^{2} \frac{f_j'(x)}{f_j(x)} \right).$$

So, it is true for $n = 1$ and $n = 2$.

Induction Step

Assume

$$\frac{d}{dx} \prod_{j=1}^{m} f_j(x) = \prod_{j=1}^{m} f_j(x) \cdot \left(\sum_{j=1}^{m} \frac{f_j'(x)}{f_j(x)} \right) \tag{88.1}$$

for all $1 \le m \le k$ for some $k \ge 2$. To prove the induction step, we need to prove

$$\frac{d}{dx} \prod_{j=1}^{k+1} f_j(x) = \prod_{j=1}^{k+1} f_j(x) \cdot \left(\sum_{j=1}^{k+1} \frac{f_j'(x)}{f_j(x)} \right).$$

By assumption (88.1) when $m = 2$,

$$\frac{d}{dx} \left(f_1(x) f_2(x) \right) = f_1(x) f_2(x) \cdot \left(\frac{f_1'(x)}{f_1(x)} + \frac{f_2'(x)}{f_2(x)} \right).$$

Let $f_1(x) \to \prod_{j=1}^{k} f_j(x)$ and $f_2(x) \to f_{k+1}(x)$,

$$\frac{d}{dx} \prod_{j=1}^{k+1} f_j(x) = \prod_{j=1}^{k+1} f_j(x) \cdot \left(\left(\prod_{j=1}^{k} f_j(x) \right)^{-1} \cdot \frac{d}{dx} \prod_{j=1}^{k} f_j(x) + \frac{f_{k+1}'(x)}{f_{k+1}(x)} \right).$$

By assumption (88.1) when $m = k$,

$$\frac{d}{dx} \prod_{j=1}^{k} f_j(x) = \prod_{j=1}^{k} f_j(x) \cdot \left(\sum_{j=1}^{k} \frac{f_j'(x)}{f_j(x)} \right).$$

So,

$$\frac{d}{dx} \prod_{j=1}^{k+1} f_j(x)$$

$$= \prod_{j=1}^{k+1} f_j(x) \cdot \left(\left(\prod_{j=1}^{k} f_j(x) \right)^{-1} \cdot \prod_{j=1}^{k} f_j(x) \cdot \left(\sum_{j=1}^{k} \frac{f_j'(x)}{f_j(x)} \right) + \frac{f_{k+1}'(x)}{f_{k+1}(x)} \right)$$

$$= \prod_{j=1}^{k+1} f_j(x) \cdot \left(\sum_{j=1}^{k} \frac{f_j'(x)}{f_j(x)} + \frac{f_{k+1}'(x)}{f_{k+1}(x)} \right)$$

$$= \prod_{j=1}^{k+1} f_j(x) \cdot \left(\sum_{j=1}^{k+1} \frac{f_j'(x)}{f_j(x)} \right).$$

Thus, we proved that it is true for $n = k+1$ if it is true for $n = m$ for all $1 \le m \le k$.

Conclusion

We proved that it is true for $n = 1$ and $n = 2$. Then, we proved that it is true for $n = k + 1$ if it is true for $n = m$ for all $1 \leq m \leq k$. Therefore, by strong induction,

$$\frac{d}{dx} \prod_{j=1}^{n} f_j(x) = \prod_{j=1}^{n} f_j(x) \cdot \left(\sum_{j=1}^{n} \frac{f_j'(x)}{f_j(x)} \right)$$

for all $n \in \mathbb{N}$.

Problem 89

Topic: Matrices

Problem: If A is an idempotent matrix, then prove that $A^n = A$ for all $n \in \mathbb{N}$.

Definition - Idempotent Matrix: An idempotent matrix is a matrix A such that $A^2 = A$.

Solution

Base Cases

We need to prove that it is true for $n = 1$. The base case $n = 1$ is trivial since it is always true that $A^1 = A$.

Induction Step

Assume that $A^k = A$ if A is an idempotent matrix for some $k \geq 1$. To prove the induction step, we need to prove that $A^{k+1} = A$ if A is an idempotent matrix. By the assumption that $A^k = A$, we have

$$A^{k+1} = AA^k = AA = A^2.$$

If A is an idempotent matrix, then by definition of idempotent matrix, $A^2 = A$. So, $A^{k+1} = A^2 = A$ if A is an idempotent matrix. Thus, we proved that it is true for $n = k + 1$ if it is true for $n = k$.

Conclusion

We proved that it is true for $n = 1$. Then, we proved that it is true for $n = k + 1$ it is true for $n = k$. Therefore, by weak induction, if A is an idempotent matrix, then $A^n = A$ for all $n \in \mathbb{N}$.

Problem 90

Topic: Matrices

Problem: Let A be an invertible matrix. Prove that the inverse of A^n is equal to the n-th power of A^{-1}, i.e. $(A^n)^{-1} = (A^{-1})^n$, for all $n \in \mathbb{N}$.

Definition - Inverse of Matrix: For an invertible matrix A, the inverse of A is the matrix A^{-1} such that

$$AA^{-1} = A^{-1}A = I,$$

where I is the identity matrix. In this problem, we will use the property

$$(AB)^{-1} = B^{-1}A^{-1}.$$

We can check that this is true by checking that

$$AB\left(B^{-1}A^{-1}\right) = \left(B^{-1}A^{-1}\right)AB = I.$$

However, note that this is NOT a rigorous proof of the property.

Solution

Base Cases

We need to prove that it is true for $n = 1$. Since

$$\left(A^1\right)^{-1} = \left(A^{-1}\right)^1 = A^{-1},$$

it is true for $n = 1$.

Induction Step

Assume

$$\left(A^k\right)^{-1} = \left(A^{-1}\right)^k \tag{90.1}$$

for some $k \geq 1$. To prove the induction step, we need to prove

$$\left(A^{k+1}\right)^{-1} = \left(A^{-1}\right)^{k+1}.$$

By using the assumption (90.1) and the property $(AB)^{-1} = B^{-1}A^{-1}$,

$$\begin{aligned}
\left(A^{k+1}\right)^{-1} &= \left(AA^k\right)^{-1} \\
&= \left(A^k\right)^{-1} A^{-1} \\
&= \left(A^{-1}\right)^k A^{-1} \\
&= \left(A^{-1}\right)^{k+1}.
\end{aligned}$$

Thus, we proved that it is true for $n = k + 1$ if it is true for $n = k$.

Conclusion

We proved that it is true for $n = 1$. Then, we proved that it is true for $n = k + 1$ if it is true for $n = k$. Therefore, by weak induction,

$$\left(A^n\right)^{-1} = \left(A^{-1}\right)^n$$

for all $n \in \mathbb{N}$.

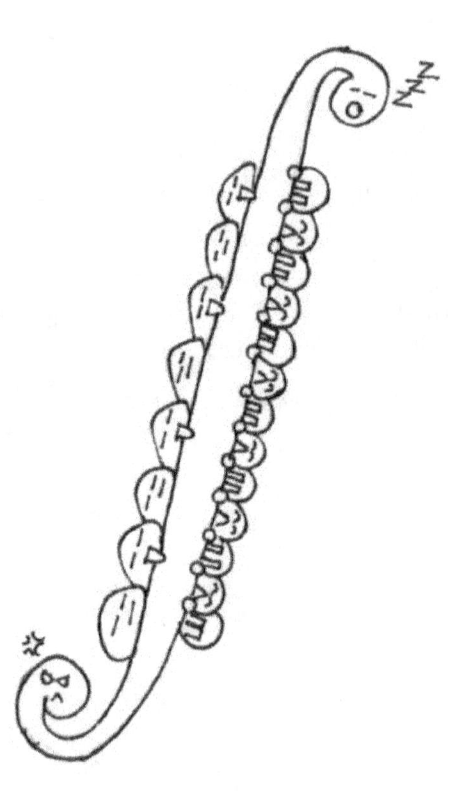

Problem 91

Topic: Matrices

Problem: Let A be a diagonalizable matrix, i.e. $A = PDP^{-1}$ where D is a diagonal matrix. Then, prove that $A^n = PD^n P^{-1}$ for all $n \in \mathbb{N}$.

Definition - Inverse of Matrix: For an invertible matrix A, the inverse of A is the matrix A^{-1} such that

$$AA^{-1} = A^{-1}A = I,$$

where I is the identity matrix.

Solution

Base Cases

We need to prove that it is true for $n = 1$. Since

$$A^1 = A = PDP^{-1} = PD^1 P^{-1},$$

it is true for $n = 1$.

Induction Step

Assume

$$A^k = PD^k P^{-1} \tag{91.1}$$

for some $k \geq 1$. To prove the induction step, we need to prove

$$A^{k+1} = PD^{k+1} P^{-1}.$$

By assumption (91.1),

$$A^{k+1} = A^k A$$
$$= \left(PD^k P^{-1}\right)\left(PDP^{-1}\right)$$
$$= PD^k \left(P^{-1}P\right)DP^{-1}$$
$$= PD^k IDP^{-1}$$
$$= PD^k DP^{-1}$$
$$= PD^{k+1}P^{-1}.$$

Thus, we proved that it is true for $n = k + 1$ if it is true for $n = k$.

Conclusion

We proved that it is true for $n = 1$. Then, we proved that it is true for $n = k + 1$ if it is true for $n = k$. Therefore, by weak induction,

$$A^n = PD^n P^{-1}$$

for all $n \in \mathbb{N}$.

Problem 92

Topic: Set Theory

Problem: If S_n is a set with n elements, then prove that S_n has 2^n subsets.

Definition - Subset: For two sets A and B, A is a subset of B, denoted as $A \subseteq B$, when all elements in A are in B. For example, $\{1, 3, 5\} \subseteq \{1, 2, 3, 5, 6\}$. Formally,

$$A \subseteq B \Leftrightarrow [\forall x, x \in A \Rightarrow x \in B].$$

Also, the empty set $\varnothing$ is a subset of any set, but this needs to be proven using the formal definition of subset. However, we will not prove this here.

Definition - Set Union: For two sets A and B, the union of A and B, denoted as $A \cup B$, is the set containing all elements in A and all elements in B. For example, $\{1, 2, 3\} \cup \{1, 3, 5\} = \{1, 2, 3, 5\}$. Formally,

$$x \in A \cup B \Leftrightarrow [x \in A \text{ or } x \in B].$$

Solution

Base Cases

We need to prove that it is true for $n = 1$. Let S_1 be a set with 1 element,

$$S_1 = \{a_1\}.$$

Then, it has 2 subsets, $\varnothing$ and $\{a_1\}$. So, it is true that a set with 1 element S_1 has subsets.

Induction Step

Assume that a k-element set S_k,

$$S_k = \{a_1, a_2, \cdots, a_k\},$$

has 2^k subsets. To prove the induction step, we need to prove that a $k+1$-element set S_{k+1},

$$S_{k+1} = \{a_1, a_2, \cdots, a_k, a_{k+1}\},$$

has 2^{k+1} subsets. Note that

$$S_{k+1} = \{a_1, a_2, \cdots, a_k\} \cup \{a_{k+1}\} = S_k \cup \{a_{k+1}\}.$$

A subset of S_{k+1} either contains or does not contain a_{k+1}.

Subsets of S_{k+1} that do not contain a_{k+1} are subsets of S_k. By the assumption that S_k has 2^k subsets, S_{k+1} has 2^k subsets that do not contain a_{k+1}.

Subsets of S_{k+1} that contain a_{k+1} are the unions of a subset of S_k with $\{a_{k+1}\}$. Since there are 2^k subsets of S_k by assumption, there are 2^k sets formed by taking the union of a subset of S_k with $\{a_{k+1}\}$. So, S_{k+1} has 2^k subsets that contain a_{k+1}.

Hence, S_{k+1} has $2^k + 2^k = 2^{k+1}$ subsets. Thus, we proved that it is true for $n = k + 1$ if it is true for $n = k$.

Conclusion

We proved that it is true for $n = 1$. Then, we proved that it is true for $n = k + 1$ if it is true for $n = k$. Therefore, by weak induction, S_n has 2^n subsets if it has n elements for all $n \in \mathbb{N}$.

Problem 93

Topic: Set Theory

Problem: Let $S_1, S_2, \cdots, S_n$ be sets such that $S_n \subseteq S_{n-1} \subseteq \cdots \subseteq S_2 \subseteq S_1$, then prove that

$$\bigcap_{i=1}^{n} S_i = S_n$$

for all $n \in \mathbb{N}$.

Definition - Set Intersection: For two sets A and B, the intersection of A and B, denoted as $A \cap B$, is the set containing all elements that are in both A and B. For example, $\{1, 2, 3\} \cap \{1, 3, 5\} = \{1, 3\}$. Formally,

$$x \in A \cap B \Leftrightarrow [x \in A \text{ and } x \in B].$$

For more than 2 sets, we have

$$\bigcap_{i=1}^{n} A_i = A_1 \cap A_2 \cap \cdots \cap A_n.$$

Definition - Subset: For two sets A and B, A is a subset of B, denoted as $A \subseteq B$, when all elements in A are in B. For example, $\{1, 3, 5\} \subseteq \{1, 2, 3, 5, 6\}$. Formally,

$$A \subseteq B \Leftrightarrow [\forall x, x \in A \Rightarrow x \in B].$$

Definition - Equal Sets: For two sets A and B, A is equal to B, denoted as $A = B$, when all elements in A are in B and all elements in B are in A. For example, $\{1, 3, 5\} = \{1, 1, 3, 5, 5\}$. Formally,

$$A = B \Leftrightarrow [\forall x, x \in A \Leftrightarrow x \in B],$$

or equivalently

$$A = B \Leftrightarrow [A \subseteq B \text{ and } B \subseteq A].$$

Solution

Base Cases

We need to prove that it is true for $n = 1$ and $n = 2$. The case when $n = 1$ is trivial since it is always true that

$$\bigcap_{i=1}^{1} S_i = S_1.$$

For $n = 2$, we need to prove that

$$\bigcap_{i=1}^{2} S_i = S_1 \cap S_2 = S_2$$

if $S_2 \subseteq S_1$.

First, we need to prove $S_1 \cap S_2 \subseteq S_2$. Let $x \in S_1 \cap S_2$. By definition of set intersection,

$$x \in S_1 \cap S_2 \Rightarrow [x \in S_1 \text{ and } x \in S_2].$$

Since the statement "$x \in S_1$ and $x \in S_2$" is true only when "$x \in S_1$" and "$x \in S_2$" are both true,

$$[x \in S_1 \text{ and } x \in S_2] \Rightarrow x \in S_2.$$

Hence,

$$\forall x, x \in S_1 \cap S_2 \Rightarrow x \in S_2,$$

and $S_1 \cap S_2 \subseteq S_2$ by definition of subset.

Next, we need to prove $S_2 \subseteq S_1 \cap S_2$. Let $x \in S_2$. If $S_2 \subseteq S_1$, then by definition of subset,

$$x \in S_2 \Rightarrow x \in S_1.$$

So,

$$x \in S_2 \Rightarrow [x \in S_1 \text{ and } x \in S_2].$$

By definition of set intersection,

$$[x \in S_1 \text{ and } x \in S_2] \Rightarrow x \in S_1 \cap S_2.$$

Hence,
$$\forall x, x \in S_2 \Rightarrow x \in S_1 \cap S_2,$$
and $S_2 \subseteq S_1 \cap S_2$ by definition of subset.

Thus, $S_1 \cap S_2 \subseteq S_2$ and $S_2 \subseteq S_1 \cap S_2$. By definition of equal sets, $S_1 \cap S_2 = S_2$ if $S_2 \subseteq S_1$.

Induction Step

Assume

$$\bigcap_{i=1}^{m} S_i = S_m \tag{93.1}$$

if $S_m \subseteq S_{m-1} \subseteq \cdots \subseteq S_2 \subseteq S_1$ for all $1 \leq m \leq k$ for some $k \geq 2$. To prove the induction step, we need to prove

$$\bigcap_{i=1}^{k+1} S_i = S_{k+1}$$

if $S_{k+1} \subseteq S_k \subseteq \cdots \subseteq S_2 \subseteq S_1$. By the assumption (93.1) when $m = k$,

$$\bigcap_{i=1}^{k} S_i = S_k$$

if $S_k \subseteq S_{k-1} \subseteq \cdots \subseteq S_2 \subseteq S_1$. So,

$$\bigcap_{i=1}^{k+1} S_i = \bigcap_{i=1}^{k} S_i \cap S_{k+1} = S_k \cap S_{k+1}.$$

By the assumption (93.1) when $m = 2$, $S_1 \cap S_2 = S_2$ if $S_2 \subseteq S_1$. Let $S_1 \to S_k$ and $S_2 \to S_{k+1}$,

$$S_k \cap S_{k+1} = S_{k+1}$$

if $S_{k+1} \subseteq S_k$. Hence,

$$\bigcap_{i=1}^{k+1} S_i = S_{k+1}$$

if $S_{k+1} \subseteq S_k \subseteq \cdots \subseteq S_2 \subseteq S_1$. Thus, we proved that it is true for $n = k+1$ if it is true for $n = m$ for all $1 \leq m \leq k$.

Conclusion

We proved that it is true for $n = 1$ and $n = 2$. Then, we proved that it is true for $n = k+1$ if it is true for $n = m$ for all $1 \leq m \leq k$. Therefore, by strong induction,

$$\bigcap_{i=1}^{n} S_i = S_n$$

$S_n \subseteq S_{n-1} \subseteq \cdots \subseteq S_2 \subseteq S_1$ for all $n \in \mathbb{N}$.

Problem 94

Topic: Set Theory

Problem: Let $S_1, S_2, \cdots, S_n$ be sets such that $S_n \subseteq S_{n-1} \subseteq \cdots \subseteq S_2 \subseteq S_1$, then prove that

$$\bigcup_{i=1}^{n} S_i = S_1$$

for all $n \in \mathbb{N}$.

Definition - Set Union: For two sets A and B, the union of A and B, denoted as $A \cup B$, is the set containing all elements in A and all elements in B. For example, $\{1, 2, 3\} \cup \{1, 3, 5\} = \{1, 2, 3, 5\}$. Formally,

$$x \in A \cup B \Leftrightarrow [x \in A \text{ or } x \in B].$$

For more than 2 sets, we have

$$\bigcup_{i=1}^{n} A_i = A_1 \cup A_2 \cup \cdots \cup A_n.$$

Definition - Subset: For two sets A and B, A is a subset of B, denoted as $A \subseteq B$, when all elements in A are in B. For example, $\{1, 3, 5\} \subseteq \{1, 2, 3, 5, 6\}$. Formally,

$$A \subseteq B \Leftrightarrow [\forall x, x \in A \Rightarrow x \in B].$$

Definition - Equal Sets: For two sets A and B, A is equal to B, denoted as $A = B$, when all elements in A are in B and all elements in B are in A. For example, $\{1, 3, 5\} = \{1, 1, 3, 5, 5\}$. Formally,

$$A = B \Leftrightarrow [\forall x, x \in A \Leftrightarrow x \in B],$$

or equivalently

$$A = B \Leftrightarrow [A \subseteq B \text{ and } B \subseteq A].$$

Solution

Base Cases

We need to prove that it is true for $n = 1$ and $n = 2$. The case when $n = 1$ is trivial since it is always true that

$$\bigcup_{i=1}^{1} S_i = S_1.$$

For $n = 2$, we need to prove that

$$\bigcup_{i=1}^{2} S_i = S_1 \cup S_2 = S_1$$

if $S_2 \subseteq S_1$.

First, we need to prove $S_1 \cup S_2 \subseteq S_1$. Let $x \in S_1 \cup S_2$. By definition of set union,

$$x \in S_1 \cup S_2 \Rightarrow [x \in S_1 \text{ or } x \in S_2].$$

If $S_2 \subseteq S_1$, then by definition of subset,

$$x \in S_2 \Rightarrow x \in S_1.$$

So,

$$[x \in S_1 \text{ or } x \in S_2] \Rightarrow [x \in S_1 \text{ or } x \in S_1] \Rightarrow x \in S_1.$$

Hence,

$$\forall x, x \in S_1 \cup S_2 \Rightarrow x \in S_1,$$

and $S_1 \cup S_2 \subseteq S_1$ by definition of subset.

Next, we need to prove $S_1 \subseteq S_1 \cup S_2$. Let $x \in S_1$. Since the statement "$x \in S_1$ or $x \in S_2$" is true when either "$x \in S_1$" or "$x \in S_2$" is true,

$$x \in S_1 \Rightarrow [x \in S_1 \text{ or } x \in S_2].$$

By definition of set union,

$$[x \in S_1 \text{ or } x \in S_2] \Rightarrow x \in S_1 \cup S_2.$$

Hence,

$$\forall x, x \in S_1 \Rightarrow x \in S_1 \cup S_2,$$

and $S_1 \subseteq S_1 \cup S_2$ by definition of subset.

Thus, $S_1 \cup S_2 \subseteq S_1$ and $S_1 \subseteq S_1 \cup S_2$. By definition of equal sets, $S_1 \cup S_2 = S_1$ if $S_2 \subseteq S_1$.

Induction Step

Assume

$$\bigcup_{i=1}^{m} S_i = S_1 \tag{94.1}$$

if $S_m \subseteq S_{m-1} \subseteq \cdots \subseteq S_2 \subseteq S_1$ for all $1 \le m \le k$ for some $k \ge 2$. To prove the induction step, we need to prove

$$\bigcup_{i=1}^{k+1} S_i = S_1$$

if $S_{k+1} \subseteq S_k \subseteq \cdots \subseteq S_2 \subseteq S_1$. By the assumption (94.1) when $m = k$,

$$\bigcup_{i=1}^{k} S_i = S_1$$

if $S_k \subseteq S_{k-1} \subseteq \cdots \subseteq S_2 \subseteq S_1$. So,

$$\bigcup_{i=1}^{k+1} S_i = \bigcup_{i=1}^{k} S_i \cup S_{k+1} = S_1 \cup S_{k+1}.$$

By the assumption (94.1) when $m = 2$, $S_1 \cup S_2 = S_1$ if $S_2 \subseteq S_1$. Let $S_2 \to S_{k+1}$,

$$S_1 \cup S_{k+1} = S_1$$

if $S_{k+1} \subseteq S_1$. Hence,

$$\bigcup_{i=1}^{k+1} S_i = S_1$$

if $S_{k+1} \subseteq S_k \subseteq \cdots \subseteq S_2 \subseteq S_1$. Thus, we proved that it is true for $n = k+1$ if it is true for $n = m$ for all $1 \le m \le k$.

Conclusion

We proved that it is true for $n = 1$ and $n = 2$. Then, we proved that it is true for $n = k+1$ if it is true for $n = m$ for all $1 \le m \le k$. Therefore, by strong induction,

$$\bigcup_{i=1}^{n} S_i = S_1$$

if $S_n \subseteq S_{n-1} \subseteq \cdots \subseteq S_2 \subseteq S_1$ for all $n \in \mathbb{N}$.

Problem 95

Topic: Integrals

Problem: Prove

$$\int_0^\infty x^n e^{-x}\,dx = n!.$$

Definition - Factorial:

$$n! = n \cdot (n-1) \cdot (n-2) \cdots 2 \cdot 1.$$

$$0! = 1.$$

Solution

Base Cases

We need to prove that it is true for $n = 1$. By integration by parts,

$$\int_0^\infty x e^{-x}\,dx = \left[-x e^{-x}\right]_0^\infty + \int_0^\infty e^{-x}\,dx$$

$$= \int_0^\infty e^{-x}\,dx$$

$$= \left[-e^{-x}\right]_0^\infty = 1 = 1!,$$

it is true for $n = 1$.

Induction Step

Assume

$$\int_0^\infty x^k e^{-x}\,dx = k! \tag{95.1}$$

for some $k \geq 1$. To prove the induction step, we need to prove

$$\int_0^\infty x^{k+1} e^{-x}\,dx = (k+1)!.$$

By integration by parts,

$$\int_0^\infty x^{k+1} e^{-x}\,dx = \left[-x^{k+1} e^{-x}\right]_0^\infty + (k+1)\int_0^\infty x^k e^{-x}\,dx$$

$$= (k+1)\int_0^\infty x^k e^{-x}\,dx.$$

Using the assumption (95.1),

$$\int_0^\infty x^{k+1} e^{-x}\,dx = (k+1) \cdot k! = (k+1)!.$$

Thus, we proved that it is true for $n = k+1$ if it is true for $n = k$.

Conclusion

We proved that it is true for $n = 1$. Then, we proved that it is true for $n = k +$ if it is true for $n = k$. Therefore, by weak induction,

$$\int_0^\infty x^n e^{-x}\,dx = n!$$

for all $n \in \mathbb{N}$.

Problem 96

Topic: Integrals, Trigonometry

Problem: Prove
$$\int_0^{\frac{\pi}{4}} \tan^{2n+1} x \, dx = \frac{(-1)^n \ln 2}{2} + \frac{(-1)^n}{2} \sum_{j=1}^{n} \frac{(-1)^j}{j}.$$

Solution

Base Cases

We need to prove that it is true for $n = 1$. By using the identity $1 + \tan^2 x = \sec^2 x$,

$$\int_0^{\frac{\pi}{4}} \tan^3 x \, dx = \int_0^{\frac{\pi}{4}} \tan x \left(\sec^2 x - 1 \right) dx$$

$$= \int_0^{\frac{\pi}{4}} \tan x \sec^2 x \, dx - \int_0^{\frac{\pi}{4}} \tan x \, dx$$

$$= \left[\frac{\tan^2 x}{2} + \ln\left(\cos x \right) \right]_0^{\frac{\pi}{4}}$$

$$= \frac{1}{2} - \frac{\ln 2}{2}$$

$$= \frac{(-1)^1 \ln 2}{2} + \frac{(-1)^1}{2} \sum_{j=1}^{1} \frac{(-1)^j}{j},$$

it is true for $n = 1$.

Induction Step

Assume

$$\int_0^{\frac{\pi}{4}} \tan^{2k+1} x \, dx = \frac{(-1)^k \ln 2}{2} + \frac{(-1)^k}{2} \sum_{j=1}^{k} \frac{(-1)^j}{j} \tag{96.1}$$

for some $k \geq 1$. To prove the induction step, we need to prove

$$\int_0^{\frac{\pi}{4}} \tan^{2k+3} x \, dx = \frac{(-1)^{k+1} \ln 2}{2} + \frac{(-1)^{k+1}}{2} \sum_{j=1}^{k+1} \frac{(-1)^j}{j}.$$

By using the identity $1 + \tan^2 x = \sec^2 x$,

$$\begin{aligned}
\int_0^{\frac{\pi}{4}} \tan^{2k+3} x \, dx &= \int_0^{\frac{\pi}{4}} \tan^{2k+1} x \left(\sec^2 x - 1 \right) dx \\
&= \int_0^{\frac{\pi}{4}} \tan^{2k+1} x \sec^2 x \, dx - \int_0^{\frac{\pi}{4}} \tan^{2k+1} x \, dx \\
&= \left[\frac{\tan^{2k+2} x}{2k+2} \right]_0^{\frac{\pi}{4}} - \int_0^{\frac{\pi}{4}} \tan^{2k+1} x \, dx \\
&= \frac{1}{2k+2} - \int_0^{\frac{\pi}{4}} \tan^{2k+1} x \, dx.
\end{aligned}$$

By the assumption (96.1),

$$\begin{aligned}
\int_0^{\frac{\pi}{4}} \tan^{2k+3} x \, dx &= \frac{1}{2k+2} - \frac{(-1)^k \ln 2}{2} - \frac{(-1)^k}{2} \sum_{j=1}^{k} \frac{(-1)^j}{j} \\
&= \frac{(-1)^{k+1}(-1)^{k+1}}{2(k+1)} + \frac{(-1)^{k+1} \ln 2}{2} + \frac{(-1)^{k+1}}{2} \sum_{j=1}^{k} \frac{(-1)^j}{j} \\
&= \frac{(-1)^{k+1} \ln 2}{2} + \frac{(-1)^{k+1}}{2} \sum_{j=1}^{k+1} \frac{(-1)^j}{j}.
\end{aligned}$$

Thus, we proved that it is true for $n = k+1$ if it is true for $n = k$.

Conclusion

We proved that it is true for $n = 1$. Then, we proved that it is true for $n = k +$
if it is true for $n = k$. Therefore, by weak induction,

$$\int_0^{\frac{\pi}{4}} \tan^{2n+1} x \, dx = \frac{(-1)^n \ln 2}{2} + \frac{(-1)^n}{2} \sum_{j=1}^{n} \frac{(-1)^j}{j}$$

for all $n \in \mathbb{N}$.

Problem 97

Topic: Integrals, Trigonometry

Problem: Prove

$$\int_0^{\frac{\pi}{2}} \sin^{2n} x \, dx = \frac{(2n)!}{4^n (n!)^2} \cdot \frac{\pi}{2}.$$

Definition - Factorial:

$$n! = n \cdot (n-1) \cdot (n-2) \cdots 2 \cdot 1.$$

$$0! = 1.$$

Solution

Base Cases

We need to prove that it is true for $n = 1$. By using the identity $\sin^2 x = \frac{1-\cos(2x)}{2}$,

$$\begin{aligned}
\int_0^{\frac{\pi}{2}} \sin^2 x \, dx &= \int_0^{\frac{\pi}{2}} \frac{1 - \cos(2x)}{2} \, dx \\
&= \left[\frac{x}{2} - \frac{\sin(2x)}{4} \right]_0^{\frac{\pi}{2}} \\
&= \frac{\pi}{4} = \frac{2!}{4^1 (1!)^2} \cdot \frac{\pi}{2},
\end{aligned}$$

it is true for $n = 1$.

Induction Step

Assume

$$\int_0^{\frac{\pi}{2}} \sin^{2k} x \, dx = \frac{(2k)!}{4^k (k!)^2} \cdot \frac{\pi}{2} \tag{97.1}$$

for some $k \geq 1$. To prove the induction step, we need to prove

$$\int_0^{\frac{\pi}{2}} \sin^{2k+2} x \, dx = \frac{(2k+2)!}{4^{k+1}((k+1)!)^2} \cdot \frac{\pi}{2}.$$

By using the identity $\sin^2 x + \cos^2 x = 1$,

$$\int_0^{\frac{\pi}{2}} \sin^{2k+2} x \, dx = \int_0^{\frac{\pi}{2}} \sin^{2k} x \left(1 - \cos^2 x\right) dx$$

$$= \int_0^{\frac{\pi}{2}} \sin^{2k} x \, dx - \int_0^{\frac{\pi}{2}} \sin^{2k} x \cos^2 x \, dx. \tag{97.2}$$

By integration by parts,

$$\int_0^{\frac{\pi}{2}} \sin^{2k} x \cos^2 x \, dx = \left[\frac{\sin^{2k+1} x}{2k+1} \cdot \cos x \right]_0^{\frac{\pi}{2}} + \frac{1}{2k+1} \int_0^{\frac{\pi}{2}} \sin^{2k+2} x \, dx$$

$$= \frac{1}{2k+1} \int_0^{\frac{\pi}{2}} \sin^{2k+2} x \, dx.$$

Substituting this into (97.2),

$$\int_0^{\frac{\pi}{2}} \sin^{2k+2} x \, dx = \int_0^{\frac{\pi}{2}} \sin^{2k} x \, dx - \frac{1}{2k+1} \int_0^{\frac{\pi}{2}} \sin^{2k+2} x \, dx$$

$$\frac{2k+2}{2k+1} \int_0^{\frac{\pi}{2}} \sin^{2k+2} x \, dx = \int_0^{\frac{\pi}{2}} \sin^{2k} x \, dx$$

$$\int_0^{\frac{\pi}{2}} \sin^{2k+2} x \, dx = \frac{2k+1}{2k+2} \int_0^{\frac{\pi}{2}} \sin^{2k} x \, dx.$$

By the assumption (97.1),

$$\int_0^{\frac{\pi}{2}} \sin^{2k+2} x \, dx = \frac{2k+1}{2k+2} \cdot \frac{(2k)!}{4^k (k!)^2} \cdot \frac{\pi}{2}$$

$$= \frac{2k+2}{2k+2} \cdot \frac{2k+1}{2k+2} \cdot \frac{(2k)!}{4^k (k!)^2} \cdot \frac{\pi}{2}$$

$$= \frac{(2k+2)!}{4^{k+1}((k+1)!)^2} \cdot \frac{\pi}{2}.$$

Thus, we proved that it is true for $n = k + 1$ if it is true for $n = k$.

Conclusion

We proved that it is true for $n = 1$. Then, we proved that it is true for $n = k + 1$ if it is true for $n = k$. Therefore, by weak induction,

$$\int_0^{\frac{\pi}{2}} \sin^{2n} x \, dx = \frac{(2n)!}{4^n (n!)^2} \cdot \frac{\pi}{2}$$

for all $n \in \mathbb{N}$.

Problem 98

Topic: Integrals, Trigonometry

Problem: Prove

$$\int_0^{\frac{\pi}{4}} \sin^{2n} x \, dx = \frac{(2n)!}{4^n (n!)^2} \left[\frac{\pi}{4} - \sum_{j=1}^{n} \frac{2^j (j!)^2}{2j(2j)!} \right].$$

Definition - Factorial:

$$n! = n \cdot (n-1) \cdot (n-2) \cdots 2 \cdot 1.$$

$$0! = 1.$$

Solution

Base Cases

We need to prove that it is true for $n = 1$. By using the identity $\sin^2 x = \frac{1-\cos(2x)}{2}$

$$\int_0^{\frac{\pi}{4}} \sin^2 x \, dx = \int_0^{\frac{\pi}{4}} \frac{1 - \cos(2x)}{2} \, dx$$

$$= \left[\frac{x}{2} - \frac{\sin(2x)}{4} \right]_0^{\frac{\pi}{4}}$$

$$= \frac{\pi}{8} - \frac{1}{4}$$

$$= \frac{2!}{4^1 (1!)^2} \left[\frac{\pi}{4} - \sum_{j=1}^{1} \frac{2^j (j!)^2}{2j(2j)!} \right],$$

so it is true for $n = 1$.

Induction Step

Assume

$$\int_0^{\frac{\pi}{4}} \sin^{2k} x \, dx = \frac{(2k)!}{4^k (k!)^2} \left[\frac{\pi}{4} - \sum_{j=1}^{k} \frac{2^j (j!)^2}{2j(2j)!} \right] \tag{98.1}$$

for some $k \geq 1$. To prove the induction step, we need to prove

$$\int_0^{\frac{\pi}{4}} \sin^{2k+2} x \, dx = \frac{(2k+2)!}{4^{k+1}((k+1)!)^2} \left[\frac{\pi}{4} - \sum_{j=1}^{k+1} \frac{2^j (j!)^2}{2j(2j)!} \right].$$

By using the identity $\sin^2 x + \cos^2 x = 1$,

$$\int_0^{\frac{\pi}{4}} \sin^{2k+2} x \, dx = \int_0^{\frac{\pi}{4}} \sin^{2k} x \left(1 - \cos^2 x \right) dx$$

$$= \int_0^{\frac{\pi}{4}} \sin^{2k} x \, dx - \int_0^{\frac{\pi}{4}} \sin^{2k} x \cos^2 x \, dx. \tag{98.2}$$

By integration by parts,

$$\int_0^{\frac{\pi}{4}} \sin^{2k} x \cos^2 x \, dx = \left[\frac{\sin^{2k+1} x}{2k + 1} \cdot \cos x \right]_0^{\frac{\pi}{4}} + \frac{1}{2k + 1} \int_0^{\frac{\pi}{4}} \sin^{2k+2} x \, dx$$

$$= \frac{1}{2^{k+1}(2k + 1)} + \frac{1}{2k + 1} \int_0^{\frac{\pi}{4}} \sin^{2k+2} x \, dx.$$

Substituting this into (98.2),

$$\int_0^{\frac{\pi}{4}} \sin^{2k+2} x \, dx = \int_0^{\frac{\pi}{4}} \sin^{2k} x \, dx - \frac{1}{2^{k+1}(2k + 1)} - \frac{1}{2k + 1} \int_0^{\frac{\pi}{4}} \sin^{2k+2} x \, dx.$$

So,

$$\frac{2k + 2}{2k + 1} \int_0^{\frac{\pi}{4}} \sin^{2k+2} x \, dx = \int_0^{\frac{\pi}{4}} \sin^{2k} x \, dx - \frac{1}{2^{k+1}(2k + 1)}$$

$$\int_0^{\frac{\pi}{4}} \sin^{2k+2} x \, dx = \frac{2k + 1}{2k + 2} \int_0^{\frac{\pi}{4}} \sin^{2k} x \, dx - \frac{1}{2^{k+2}(k + 1)}.$$

By the assumption (98.1),

$$\int_0^{\frac{\pi}{4}} \sin^{2k+2} x \, dx = \frac{2k+1}{2k+2} \cdot \frac{(2k)!}{4^k (k!)^2} \left[\frac{\pi}{4} - \sum_{j=1}^{k} \frac{2^j (j!)^2}{2j(2j)!} \right] - \frac{1}{2^{k+2}(k+1)}$$

$$= \frac{(2k+2)!}{4^{k+1}((k+1)!)^2} \left[\frac{\pi}{4} - \sum_{j=1}^{k} \frac{2^j (j!)^2}{2j(2j)!} \right] - \frac{1}{2^{k+2}(k+1)}$$

$$= \frac{(2k+2)!}{4^{k+1}((k+1)!)^2} \left[\frac{\pi}{4} - \sum_{j=1}^{k+1} \frac{2^j (j!)^2}{2j(2j)!} \right].$$

Thus, we proved that it is true for $n = k+1$ if it is true for $n = k$.

Conclusion

We proved that it is true for $n = 1$. Then, we proved that it is true for $n = k+1$ if it is true for $n = k$. Therefore, by weak induction,

$$\int_0^{\frac{\pi}{4}} \sin^{2n} x \, dx = \frac{(2n)!}{4^n (n!)^2} \left[\frac{\pi}{4} - \sum_{j=1}^{n} \frac{2^j (j!)^2}{2j(2j)!} \right]$$

for all $n \in \mathbb{N}$.

Problem 99

Topic: Integrals, Trigonometry

Problem: Prove

$$\int_0^{\frac{\pi}{4}} \sec^{2n} x \, dx = \frac{2^{2n-1}(n!)^2}{n\,(2n)!} \sum_{j=0}^{n-1} \frac{(2j)!}{2^j\,(j!)^2}.$$

Definition - Factorial:

$$n! = n \cdot (n-1) \cdot (n-2) \cdots 2 \cdot 1.$$

$$0! = 1.$$

Solution

Base Cases

We need to prove that it is true for $n = 1$. Since

$$\int_0^{\frac{\pi}{4}} \sec^2 x \, dx = [\tan x]_0^{\frac{\pi}{4}} = 1 = \frac{2^{2\cdot 1-1}(1!)^2}{1 \cdot 2!} \sum_{j=0}^{1-1} \frac{(2j)!}{2^j\,(j!)^2},$$

is true for $n = 1$.

Induction Step

Assume

$$\int_0^{\frac{\pi}{4}} \sec^{2k} x \, dx = \frac{2^{2k-1}(k!)^2}{k\,(2k)!} \sum_{j=0}^{k-1} \frac{(2j)!}{2^j\,(j!)^2} \tag{99.1}$$

for some $k \geq 1$. To prove the induction step, we need to prove

$$\int_0^{\frac{\pi}{4}} \sec^{2k+2} x \, dx = \frac{2^{2k+1}((k+1)!)^2}{(k+1)(2k+2)!} \sum_{j=0}^{k} \frac{(2j)!}{2^j (j!)^2}.$$

By integration by parts and using the identity $1 + \tan^2 x = \sec^2 x$,

$$\int_0^{\frac{\pi}{4}} \sec^{2k+2} x \, dx = \int_0^{\frac{\pi}{4}} \sec^{2k} x \sec^2 x \, dx$$

$$= \left[\sec^{2k} x \tan x \right]_0^{\frac{\pi}{4}} - 2k \int_0^{\frac{\pi}{4}} \sec^{2k} x \tan^2 x \, dx$$

$$= 2^k - 2k \int_0^{\frac{\pi}{4}} \sec^{2k} x \left(\sec^2 x - 1 \right) dx$$

$$= 2^k - 2k \int_0^{\frac{\pi}{4}} \sec^{2k+2} x \, dx + 2k \int_0^{\frac{\pi}{4}} \sec^{2k} x \, dx.$$

So,

$$\int_0^{\frac{\pi}{4}} \sec^{2k+2} x \, dx = \frac{2^k}{2k+1} + \frac{2k}{2k+1} \int_0^{\frac{\pi}{4}} \sec^{2k} x \, dx.$$

By the assumption (99.1),

$$\int_0^{\frac{\pi}{4}} \sec^{2k+2} x \, dx = \frac{2^k}{2k+1} + \frac{2k}{2k+1} \cdot \frac{2^{2k-1}(k!)^2}{k(2k)!} \sum_{j=0}^{k-1} \frac{(2j)!}{2^j (j!)^2}$$

$$= \frac{2^k}{2k+1} + \frac{2^{2k+1}((k+1)!)^2}{(k+1)(2k+2)!} \sum_{j=0}^{k-1} \frac{(2j)!}{2^j (j!)^2}$$

$$= \frac{2^{2k+1}((k+1)!)^2}{(k+1)(2k+2)!} \sum_{j=0}^{k} \frac{(2j)!}{2^j (j!)^2}.$$

Thus, we proved that it is true for $n = k + 1$ if it is true for $n = k$.

Conclusion

We proved that it is true for $n = 1$. Then, we proved that it is true for $n = k + $
if it is true for $n = k$. Therefore, by weak induction,

$$\int_0^{\frac{\pi}{4}} \sec^{2n} x \, dx = \frac{2^{2n-1}(n!)^2}{n(2n)!} \sum_{j=0}^{n-1} \frac{(2j)!}{2^j (j!)^2}$$

for all $n \in \mathbb{N}$.

Problem 100

Topic: Integrals, Trigonometry

Problem: Prove

$$\int_0^{\frac{\pi}{2}} x^{2n} \cos x \, dx = \sum_{j=0}^{n} \frac{(-1)^n (2n)!}{(-1)^j (2j)!} \left(\frac{\pi}{2}\right)^{2j}.$$

Definition - Factorial:

$$n! = n \cdot (n-1) \cdot (n-2) \cdots 2 \cdot 1.$$

$$0! = 1.$$

Solution

Base Cases

We need to prove that it is true for $n = 1$. By integration by parts twice,

$$\int_0^{\frac{\pi}{2}} x^2 \cos x \, dx = \left[x^2 \sin x\right]_0^{\frac{\pi}{2}} - 2 \int_0^{\frac{\pi}{2}} x \sin x \, dx$$

$$= \left(\frac{\pi}{2}\right)^2 - 2 \int_0^{\frac{\pi}{2}} x \sin x \, dx$$

$$= \left(\frac{\pi}{2}\right)^2 + \left[2x \cos x\right]_0^{\frac{\pi}{2}} - 2 \int_0^{\frac{\pi}{2}} \cos x \, dx$$

$$= \left(\frac{\pi}{2}\right)^2 - 2 \int_0^{\frac{\pi}{2}} \cos x \, dx = \left(\frac{\pi}{2}\right)^2 - 2.$$

Since

$$\sum_{j=0}^{1} \frac{(-1)^1\, 2!}{(-1)^j (2j)!} \left(\frac{\pi}{2}\right)^{2j} = \left(\frac{\pi}{2}\right)^2 - 2,$$

it is true that

$$\int_0^{\frac{\pi}{2}} x^2 \cos x\, dx = \sum_{j=0}^{1} \frac{(-1)^1\, 2!}{(-1)^j (2j)!} \left(\frac{\pi}{2}\right)^{2j}.$$

Induction Step

Assume

$$\int_0^{\frac{\pi}{2}} x^{2k} \cos x\, dx = \sum_{j=0}^{k} \frac{(-1)^k (2k)!}{(-1)^j (2j)!} \left(\frac{\pi}{2}\right)^{2j} \qquad (100.1)$$

for some $k \geq 1$. To prove the induction step, we need to prove

$$\int_0^{\frac{\pi}{2}} x^{2k+2} \cos x\, dx = \sum_{j=0}^{k+1} \frac{(-1)^{k+1}(2k+2)!}{(-1)^j (2j)!} \left(\frac{\pi}{2}\right)^{2j}.$$

By integration by parts twice,

$$\int_0^{\frac{\pi}{2}} x^{2k+2} \cos x\, dx$$

$$= \left[x^{2k+2} \sin x \right]_0^{\frac{\pi}{2}} - (2k+2) \int_0^{\frac{\pi}{2}} x^{2k+1} \sin x\, dx$$

$$= \left(\frac{\pi}{2}\right)^{2k+2} - (2k+2) \int_0^{\frac{\pi}{2}} x^{2k+1} \sin x\, dx$$

$$= \left(\frac{\pi}{2}\right)^{2k+2} + (2k+2) \left[x^{2k+1} \cos x \right]_0^{\frac{\pi}{2}} - (2k+2)(2k+1) \int_0^{\frac{\pi}{2}} x^{2k} \cos x\, dx$$

$$= \left(\frac{\pi}{2}\right)^{2k+2} - (2k+2)(2k+1) \int_0^{\frac{\pi}{2}} x^{2k} \cos x\, dx.$$

By the assumption (100.1),

$$\int_0^{\frac{\pi}{2}} x^{2k+2} \cos x\, dx = \left(\frac{\pi}{2}\right)^{2k+2} - (2k+2)(2k+1) \sum_{j=0}^{k} \frac{(-1)^k (2k)!}{(-1)^j (2j)!} \left(\frac{\pi}{2}\right)^{2j}$$

$$= \left(\frac{\pi}{2}\right)^{2k+2} + \sum_{j=0}^{k} \frac{(-1)^{k+1}(2k+2)!}{(-1)^j (2j)!} \left(\frac{\pi}{2}\right)^{2j}$$

$$= \sum_{j=0}^{k+1} \frac{(-1)^{k+1}(2k+2)!}{(-1)^j (2j)!} \left(\frac{\pi}{2}\right)^{2j}.$$

Thus, we proved that it is true for $n = k+1$ if it is true for $n = k$.

Conclusion

We proved that it is true for $n = 1$. Then, we proved that it is true for $n = k + 1$ if it is true for $n = k$. Therefore, by weak induction,

$$\int_0^{\frac{\pi}{2}} x^{2n} \cos x \, dx = \sum_{j=0}^{n} \frac{(-1)^n (2n)!}{(-1)^j (2j)!} \left(\frac{\pi}{2}\right)^{2j}$$

for all $n \in \mathbb{N}$.

Problem 101

Topic: Integrals, Trigonometry

Problem: Prove

$$\int_0^{\frac{\pi}{2}} x \cos^{2n} x \, dx = \frac{(2n)!}{4^n (n!)^2} \left[\frac{\pi^2}{8} - \sum_{j=0}^{n-1} \frac{4^j (j!)^2}{(2j+2)!} \right].$$

Definition - Factorial:

$$n! = n \cdot (n-1) \cdot (n-2) \cdots 2 \cdot 1.$$

$$0! = 1.$$

Solution

Base Cases

We need to prove that it is true for $n = 1$. By using the identity $\cos^2 x = \frac{1+\cos(2x)}{2}$

$$\int_0^{\frac{\pi}{2}} x \cos^2 x \, dx = \int_0^{\frac{\pi}{2}} \frac{x + x \cos(2x)}{2} \, dx$$

$$= \left[\frac{x^2}{4} + \frac{x \sin(2x)}{4} + \frac{\cos(2x)}{8} \right]_0^{\frac{\pi}{2}}$$

$$= \frac{\pi^2}{16} - \frac{1}{4}$$

$$= \frac{2!}{4^1 (1!)^2} \left[\frac{\pi^2}{8} - \sum_{j=0}^{1-1} \frac{4^j (j!)^2}{(2j+2)!} \right],$$

so it is true for $n = 1$.

Induction Step

Assume

$$\int_0^{\frac{\pi}{2}} x \cos^{2k} x \, dx = \frac{(2k)!}{4^k (k!)^2} \left[\frac{\pi^2}{8} - \sum_{j=0}^{k-1} \frac{4^j (j!)^2}{(2j+2)!} \right] \tag{101.1}$$

for some $k \geq 1$. To prove the induction step, we need to prove

$$\int_0^{\frac{\pi}{2}} x \cos^{2k+2} x \, dx = \frac{(2k+2)!}{4^{k+1}((k+1)!)^2} \left[\frac{\pi^2}{8} - \sum_{j=0}^{k} \frac{4^j (j!)^2}{(2j+2)!} \right].$$

By using the identity $\sin^2 x + \cos^2 x = 1$,

$$\int_0^{\frac{\pi}{2}} x \cos^{2k+2} x \, dx = \int_0^{\frac{\pi}{2}} x \cos^{2k} x \left(1 - \sin^2 x \right) dx$$

$$= \int_0^{\frac{\pi}{2}} x \cos^{2k} x \, dx - \int_0^{\frac{\pi}{2}} x \cos^{2k} x \sin^2 x \, dx. \tag{101.2}$$

By integration by parts,

$$- \int_0^{\frac{\pi}{2}} x \cos^{2k} x \sin^2 x \, dx$$

$$= \int_0^{\frac{\pi}{2}} (x \sin x) \left(- \cos^{2k} x \sin x \right) dx$$

$$= \left[x \sin x \cdot \frac{\cos^{2k+1} x}{2k+1} \right]_0^{\frac{\pi}{2}} - \frac{1}{2k+1} \int_0^{\frac{\pi}{2}} (\sin x + x \cos x) \cos^{2k+1} x \, dx$$

$$= - \frac{1}{2k+1} \int_0^{\frac{\pi}{2}} \sin x \cos^{2k+1} x \, dx - \frac{1}{2k+1} \int_0^{\frac{\pi}{2}} x \cos^{2k+2} x \, dx$$

$$= - \frac{1}{(2k+2)(2k+1)} - \frac{1}{2k+1} \int_0^{\frac{\pi}{2}} x \cos^{2k+2} x \, dx.$$

Substituting this into (101.2) and simplifying algebraically gives

$$\int_0^{\frac{\pi}{2}} x \cos^{2k+2} x \, dx = \frac{2k+1}{2k+2} \int_0^{\frac{\pi}{2}} x \cos^{2k} x \, dx - \frac{1}{4(k+1)^2}.$$

By the assumption (101.1),

$$\int_0^{\frac{\pi}{2}} x \cos^{2k+2} x \, dx = \frac{2k+1}{2k+2} \cdot \frac{(2k)!}{4^k (k!)^2} \left[\frac{\pi^2}{8} - \sum_{j=0}^{k-1} \frac{4^j (j!)^2}{(2j+2)!} \right] - \frac{1}{4(k+1)^2}$$

$$= \frac{(2k+2)!}{4^{k+1}((k+1)!)^2} \left[\frac{\pi^2}{8} - \sum_{j=0}^{k-1} \frac{4^j (j!)^2}{(2j+2)!} \right] - \frac{1}{4(k+1)^2}$$

$$= \frac{(2k+2)!}{4^{k+1}((k+1)!)^2} \left[\frac{\pi^2}{8} - \sum_{j=0}^{k} \frac{4^j (j!)^2}{(2j+2)!} \right].$$

Thus, we proved that it is true for $n = k+1$ if it is true for $n = k$.

Conclusion

We proved that it is true for $n = 1$. Then, we proved that it is true for $n = k+1$ if it is true for $n = k$. Therefore, by weak induction,

$$\int_0^{\frac{\pi}{2}} x \cos^{2n} x \, dx = \frac{(2n)!}{4^n (n!)^2} \left[\frac{\pi^2}{8} - \sum_{j=0}^{n-1} \frac{4^j (j!)^2}{(2j+2)!} \right]$$

for all $n \in \mathbb{N}$.

Acknowledgements

In this section, I would like to mention some people whom I would like to thank
for their help during the process of producing this book.

On Instagram, I have established an edit team consisting of a few members with
various math backgrounds. I would like to thank those members for their helpful
feedback and suggestions for the book. The following is the list of the members of
the edit team in no particular order:

- Thang Pang Ern

- Keyvon Rashidi

- Mislav Plavac

I also appreciate Andrzej Kukla, the cover designer of this book, and Vy Nguyen
Tong, the illustrator of this book, for their artistic contribution to make this book
more than just a boring math book with a simple cover and a long list of problems

Reviews

"In the ideal situation, every college student should learn some mathematics, with the depth and difficulty largely varying. As in the practice of most refined universities, mathematical education should never be standardized. Meanwhile, math induction is one of the key tools that students need to know. Based on my teaching experiences to many Chinese and international students, I found Tran's book fairly interesting and accessible.

Mathematical Induction 101 contains 101 carefully chosen exercise problems. I believe that Tran's book will prove to be useful for college students in the broad range of arts and sciences. Hopefully many will appreciate the beauty and power of mathematical reasoning with mastery of math induction after using this book."

Dr. Chunwei Song
Professor of Mathematics, Peking University

"This book provides an interesting collection of identities and inequalities that can be proved by mathematical induction. These 101 practices will be useful to learn standard techniques. Some of the topics and their solutions would also be found stimulating even for advanced students of mathematics."

Dr. Hiroaki Nakamura
Professor of Mathematics, Osaka University

"This book provides almost all types of problems in high school and collegiate mathematics which can be solved by mathematical induction. To each problem, a self-contained and detailed solution is given. Difficulty ranges from quite elementary to somewhat complex. Among them there are many interesting ones. I am particularly interested in problems related to the Fibonacci sequence, floor function, and iterated functions."

Dr. Young-One Kim
Professor Emeritus of Mathematics, Seoul National University

"As a mathematics student, this is a book I would have loved in my introductory proof class. Duc Van Khanh Tran has written up a fantastic way to both learn and practice one of the most essential tools in a mathematician's toolkit: Induction!

It's difficult to find such a large, comprehensive array of well-written and demonstrative practice problems, which makes this all the more valuable a resource. Duc Van Khanh Tran carefully pens each solution to be as intuitive and explanatory as possible, so the risk of confusion at a solution is minimal.

The vast amount of problems is complemented by the vast array of topics the book covers. Inequalities, sequences, trigonometry, power series, derivatives, and integrals are only some of the concepts found, guaranteeing that anyone interested in math will get something new out of this book. I've tried a few of these for fun! I really like the integral problems myself.

All in all, if you'd like to learn about induction, hone your induction skills, or just want a bunch of fun induction problems, purchase away! You won't regret it!"

Said Kaili

Undergraduate Student of Mathematics, University of Virginia

"Despite minimal prior proof background, I can grasp the main scopes of the book without much difficulty. Indeed, Duc Tran's book provides clear instructions, comprehensive content, and helpful problems that ease students into learning mathematical induction."

Ky Minh Vinh Nguyen

Undergraduate Student of Informatics, University of Texas at Austin